GENETIC MAPS
AND
HUMAN
IMAGINATIONS

GENETIC MAPS
AND
HUMAN IMAGINATIONS

The Limits of Science in
Understanding Who We Are

BARBARA KATZ ROTHMAN

W. W. NORTON & COMPANY
New York · London

For information about permission to reproduce selections from this book,
write to permissions, W. W. Norton & Company, Inc.,
500 Fifth Avenue, New York, NY 10110.

The text and display of this book are composed in Opti Cuba Libre Two
Composition and manufacturing by the Haddon Craftsmen, Inc.
Book design by BTD/Mary A. Wirth

LIBRARY OF CONGRESS CATALOGING-IN-PUBLICATION DATA

Rothman, Barbara Katz.
Genetic maps and human imaginations :
the limits of science in understanding who we are /
Barbara Katz Rothman.
p.   cm.
includes bibliographical references and index.
ISBN 0-393-04703-2
1. Human genetics—Philosophy.
2. Human genome—Research—Moral and ethical aspects.
3. Human beings—Philosophy.
I. Title.
QH431.R8526   1998   98-18800
599.93'5'01—dc21   CIP

W. W. Norton & Company, Inc.
500 Fifth Avenue
New York, N.Y. 10110
http://www.wwnorton.com

W. W. Norton & Company Ltd.
10 Coptic Street
London WC1A 1PU

1   2   3   4   5   6   7   8   9   0

For Joan Colb Fried, in loving memory

Contents

7

IMAGINING THE FUTURE:
THE MICROEUGENICS OF PROCREATION

CONCLUSION

GENETIC MAPS
AND
HUMAN
IMAGINATIONS

INTRODUCTION

Before I Start

Genetics isn't just a science. It's becoming more than that. It's a way of thinking, an ideology. We're coming to see life through a "prism of heritability," a "discourse of gene action," a genetics frame. Genetics is the single best explanation, the most comprehensive theory since God. Whatever the question is, genetics is the answer. Every possible issue of our time—race and racism, addictions, war, cancer, sexuality—all of it has been placed in the genetics frame.

Genetics is the contemporary frontier in science. It is the place we now look to when we want to understand the big questions: our place in the cosmos, the meaning of life. Not terribly long ago we thought the answer might lie "out there," in space; now we look deep inside, into the nucleus of the cell. One of the ways of understanding human history is to look at the ways that we have asked and sought to answer the most fundamental questions. Our knowledge and our quest for knowledge reflect our historical moment.

All knowledge is knowledge from somewhere. Every way of knowing the world is grounded, placed, located. When I'm trying to explain that to a hundred business majors in an Introduction to Sociology class at 7:30 in the morning, I end up using props, and dancing around the table in front of the room. I line up my briefcase, a notebook, a pen. If you look from one side of the table, the briefcase blocks the notebook—it might as well not be there. Run over to the other side of the table and you see all three objects, but the briefcase is just a big black lump, its zippers and flaps turned away, hidden. Look down from above and the notebook is the center of everything, and you still can't see the dark side of the briefcase. There is nowhere you can stand and see it all, and anywhere we stand puts things in relationship to one another: in front of, behind, close, far, first, last.

Genetics as a science is grounded, placed, located in its moment in history. As am I, a middle-aged white woman who has borne two children and raised three, studied sociology, participated and failed to participate in the history of her time.

The new genetics and its body of knowledge are grounded in the political context of our time. It is a context in which we have been moving to the political right, in which intellectually respectable racism raised its head yet again, in which the starvation of children has been seen as justifiable punishment for wayward parents, in which "family values" seem to be more highly valued than have actual families. This is the context in which the new genetics research has taken place. Is the new genetics research about knowledge for its own sake? Every possible area of basic research has been cut, and cut again, and cut yet again. But genetics gets funded. Why is that? Is the new genetics about health and medicine, helping us to live longer and healthier lives? Simply to say "health care" in the United States today is to hear the echo of cost containment, budget slashings, rationing. But genetics gets funded. And doesn't that make you wonder?

Science doesn't occur in a political or a social vacuum. Many things have changed over the last thirty years, including the direction of scientific work. Somehow, when we were just at the brink of understanding the connectedness of the world, when people were just starting to understand the concept of ecological systems, of the link between healthy environments and healthy people, at that exact moment we found ourselves backing away, looking inward.

The influence of genetics is everywhere: the gene is an icon of our time. The news gives us one genetics headline after another. Genes have replaced hormones as the sometimes humorous, sometimes serious "excuse mode" in everything from jokes to novels to courtrooms. A new introductory sociology text uses the double helix as its cover design. Everywhere we turn, throughout our language, our arts and our sciences, genetic thinking dominates.

The deep logic of genetics, the frame or prism of understanding that genetics gives us, is that genes are *causes*. As Evelyn Fox Keller, philosopher of science, summarizes the logic of genetic thinking, "genes are the primary agents of life; they are the fundamental units of biological analysis; they cause the development of biological traits; and the ultimate goal of biological science is the understanding of how they act." If genes are the cause, the active force, the predictor of traits, then to read genes is to predict traits.

But geneticists often cannot predict traits. There is a murkiness in the translation, not the direct cause and effect the logic suggests. And so geneticists have introduced a useful distinction, noting the difference between genotype and phenotype. The genotype is the genetic reading. The phenotype is how it actually played out, the being before you. "Environment" can then be understood as that which muddies the waters, that which interferes between the genotype and the phenotype. In a particular pregnancy, for example, the poorly placed identical twin will be smaller than the more felicitously implanted twin; their different environments within the same womb account for their phenotypical difference in spite of their genotypical identity.

Something funny happens to our thinking with this model. The genotype comes to seem more real somehow, more authentic, what was meant to be. Better. The smaller twin was stunted, we tend to think, rather than that the larger was very well nurtured. All through genetic determinist thinking the environment—at the level of the cytoplasm of the rest of the cell, outside of the DNA-laden nucleus; at the level of the body that carries that embryo; of the community that shelters that maternal body; of the society that contains that community; of the planet on which we all live—all of the environment comes to be seen as that which might interfere with the expression of the genotype that was meant to be.

Genetics—a way of thinking, an ideology as much as a science—puts all of the essence of life, all of its energy, majesty and power, into the nucleus of the cell. The old-fashioned word for that essential bit, that source of life, was the "seed." In the history of Western society, that seed was something that men had. Our standard "where do babies come from?" tale is a variation on Daddy plants a seed in Mommy.

I come to the study of genetics with my own history, my own grounding, place, location. I am a mother and I am a sociologist who has studied motherhood. My professional, intellectual commitment and my personal, moral commitment are to the fabric of life, the connections between people. Seeds don't fascinate me. Relationships do. Motherhood fascinates me—its physical, bodily connectedness as well as its social, caring connectedness. My favorite children's book about where babies come from is by Pearl Buck. Johnny wants to know where he came from. You were in me, his mother explains. But *before*, Johnny wants to know. You were always in me, his mother explains. When I was in my mother you were in me. When she was in her mother, you were in me. You were always in me.

I don't want to substitute women's seeds for men's seeds; that's not the point. The point is, rather, that we unfold out of each other's bodies, that we're not planted by something, by anything, but that life begets life, that we grow from and of and within each other.

Modern thinking has recognized women's seeds, and has extended some of the privileges of fatherhood to women. We no longer talk of women as just vessels for the children of men. Women are recognized as being connected to their children, like men are, through their seeds. Children, we say, are "half his, half hers"—as if they might as well have grown in the backyard.

In this seed-based way of thinking, in its most primitive forms and in the most sophisticated genetics, when people talk about biological connectedness, they are talking about a genetic tie, a connection by seed. We use the word "blood," an older language of bodily connectedness, and talk about "blood ties." But we mean "genetic ties," the seed connection: the one absolutely bloodless part of making babies. Children do grow out of the blood of their mothers, of their bodies and being. The mother-based tie is the growing of children, the carrying and bearing and raising; the patriarchal tie is based on genetics, the seed connection.

Each of these ways of thinking leads to different ideas about what a person is. In the matriarchal societies of the world, a person is what mothers grow. People are made of the care and nurturance that brings a baby forth into the world and turns that baby into a member of the society. In patriarchies, a person is what grows out of a seed. The essence of the person, who the person really is, is there in the seed when it is planted in the mother. Early scientists in Western society were so deeply committed to the father-based way of thinking that it influenced what they saw. One of the first uses of the microscope was to look at semen and see the little person, the "homunculus," curled up inside the sperm. And the director of a California sperm bank distributes t-shirts with a drawing of sperm swimming on a blue background, accompanied by the words "Future People."

But people don't begin as separate beings, despite this imagery of sperm on a blank background. We begin as part of our mothers' bodies. Our bodies grow out of the bodies that surround us. We don't, as our language would have us believe, "enter the world," or "arrive." From where? Women who give birth, I have often pointed out, don't feel babies *arrive*. We feel them *leave*.

People do not "spring up like mushrooms," as Thomas Hobbes and the other social contract theorists we studied in high school civics would have us believe. We are conceived inside of bodies, we come forth after months of hearing voices, feeling the rhythm of the human body, cradled in the pelvic rock of our mothers' walk. We move from inside the body to outside. Right outside. We spend years in intimate physical contact with other bodies, being cradled, held, rocked, carried, suckled. That is who we are and how we got to be who we are, not separate beings that must learn to cope with others, but attached beings that must learn how to separate. Cradle a child, sit by the side of someone sick or frightened or in pain—we cling to one another. For a few moments now and again, when everything is going okay, it is possible to hold onto the illusion that we are separate individuals. But our connectedness is the reality; the separation is the illusion.

The sociological tradition in which I am grounded has taught

me to see mind, knowledge, and understanding as social phenom-
ena. In a way, each mind is just a refraction, a prism, through which
one sees *the* mind, the social consciousness, the social world. We
speak our own thoughts, but we speak them in our shared language.

I am making much the same claim about the body. Each individual is part of the whole. We unfold from inside one another. Pregnancy for me is not a metaphor of connectedness. It *is* connectedness.
I have given birth and I have been beside women at their births. I have
held placentas in my hand. I have felt pulsing cords. I know we are
born connected. Each body, each individual, is part of the whole, the
temporary form that human life takes in that time and space, a reflection or refraction of others. The mind I experience as my very own is
my take on the collective social history and experience of my world;
my body is the form that this particular dip in the gene pool has
taken.

Genetics isn't just a science; it is a way of thinking. Ruth Hubbard, professor emerita at Harvard and herself a biologist, points out
that scientists are not detached observers of nature. They are con-
stantly making decisions about what they will consider significant,
and these choices are not merely individual or idiosyncratic, but reflect the society in which the scientists live and work. Genetic thinking, genetics as our society is developing it, genetics as we use it and
think it and speak it, genetics as an ideology for our time, is about
seeds. Genetics is the most obvious and direct scientific descendent
of this traditional, father-based worldview, an attempt to understand the very meaning of life by understanding genes, bits of DNA,
the updated version of seeds.

The Human Genome Project, the multi-billion-dollar international attempt to map the human genome (the distribution of genes
on chromosomes), to find where each gene lies and ultimately what
each does, is frequently described by its critics as the ultimate reductionist project. And reductionist it surely is. But to define it as such
is in a way to accept the underlying premise that we are reducible to
our genes. Or, as my colleague Alan Spector has put it, the soul has
moved into the genes. The Human Genome Project is an attempt to

reduce us to biology, to go back to what are thought of as first causes. Genetics attempts to explain people, our physical and our social presence, by going back to the seed, the moment of zygotic zero, when sperm joins egg. Spector is not wrong when he speaks of the soul: in a secular society, this is as close as we come to a moment of ensoulment.

In this way of thinking, the seed contains all it could be. It is pure potential. Everything else becomes background. Look at a green field. A mother-based worldview sees the rich green earth springing forth with life. We, with our genetics thinking, see seeds put into dirt, into soil. Think about those words we use for the very earth: dirt, soil. That which is precious and life-giving has moved into the seed; the Earth itself is but dirt. Donna King, a sociologist who has studied children's environmental literature, has pointed out that in books like *The Lorax* and films like *Ferngully,* the planet will be saved by saving the precious seed.

But is saving the seed our real problem? The threat to the land, water, air, to "Mother Earth" is our problem. If the seeds of the Earth *are* in danger, genetic engineering is more the cause than the solution. Diversity is being bred out; the square tomato bred in.

My own concerns are less at the level of the ecological system and more at the level of the social system. The social world, like the Earth, needs our protection and our nurturance. Social problems lie in the social world, not in the quality of the individual inhabitants of that world. I don't want the people who are bringing us a more packagable, storable, cost-efficient tomato turning their attention to my grandchildren. We talk about the perfect baby, and how genetics can help us achieve that. But perfection is *for* something. The perfect tomato is perfectly marketable. And the perfect baby?

What is it we are trying to do? What is the point? More than a decade ago, I wrote a book on women's experiences with prenatal diagnosis, the cutting edge of applied genetic technology at the time. I interviewed women who had had prenatal testing and women who had refused it, and women who had gotten bad news and ended their pregnancies after the diagnosis. It was a wrenching, heartbreaking research project. But intellectually, it was fascinating. It brought me into the world of genetics, of bioethics. And finally, after trying to figure it all out—what it means to have a baby, to use the testing to avoid having a particular baby, to end a pregnancy—finally I realized that there was only one more thing I needed to know. The missing piece was the meaning of life.

I could figure out the Human Genome Project, what it means to "map" all of our genes onto their spots on the chromosomes, to read that, to compare genomes between groups of people and potentially even to make changes in those genes. I could figure out what it means and what to do about it, if only I could get this one missing bit filled in: What is the meaning of life? What is the point?

Mapping the genome won't answer that question. It makes the answer urgent, but it does not and cannot answer the question. Some people feel they have the answer. For some it is answered in organized religious belief. Some find the answer in a more idiosyncratic spirituality. I personally am at a loss. I'm spiritually tone deaf, don't resonate somehow to religious answers. But I've still got questions, and I do resonate to them.

On Breads, Bibles and Blueprints

I'm on the subway; at the beach; in an elevator; on a family vacation in Miami; talking to a student; listening to the radio. And I hear, "It's genetic." Oftentimes it is followed by laughter. "It's genetic" is offered as an excuse, a kind of throwing up of one's hands, helpless before a larger force. Someone lost her keys; missed her stop; got an A in spelling; lived to 97; can't lose weight; can't gain weight; got an F in spelling; remembered every phone number she's ever learned.

"It's genetic."

It's a long way from that to what geneticists are calling genetic.

The word "gene" preceded the current biological understanding of genetics. "Gene" was the name given to the force that transmits qualities from parent to child, whether among people or among pea plants. What is *now* called "genes" are stretches of DNA that code for the production of specific proteins.

DNA is that famous double helix, that intertwining spiral staircase whose image shows up on everything from ads to sociology textbooks. There are four molecules that make up DNA: adenine, thymine, cytosine, and guanine, usually identified only as A, T, C, G. Each molecule is in the form of a disc, called a "base," and the rungs of the ladder that connects the two spirals are made when two molecular discs form what are called "base pairs." Adenine pairs with thymine; cytosine pairs with guanine. The human genome is made up of these base pairs, about three billion of them, arranged on 23 pairs of chromosomes. Eyes glaze over every time people read this: truth is, you probably don't need to know the names of these bases or remember which pairs with which.

But what it *does* help to remember is what the DNA does. About

10 percent of DNA, segments ranging from as few as several thousand base pairs to as many as several hundred thousand base pairs, are called genes. Each of these segments of DNA "codes" for the production of a specific protein. Proteins are themselves made up of amino acids in a particular sequence: three base pairs translates into one amino acid.

Now this begins to make some sense: if a person (or a banana plant, for that matter) is missing a correct sequence of DNA in one of these segments, they won't make some crucial protein. Hemophilia is a good example: there is a protein that is needed for blood to clot properly, and some people have errors in the relevant stretch of DNA and do not produce that protein. It's still not a simple situation: in a study of 216 people with hemophilia B, the mutations or errors occurred in 115 different places on the gene. But still, that example makes sense: if you inherit a gene with an error, you won't produce some protein, and you might suffer unpleasant consequences as a result.

What is a little harder to follow is how you go from coding for protein production to who in a family has big ears, or dimples, or a crooked middle toe—all things we understand as fairly simply "genetic" in origin. Predicting more complex things—like a tendency to misplace the car keys or to remember phone numbers—requires seeing a connection between a protein and a way of behaving or thinking.

A few years back, before everything was "genetic," the standard explanation was "it's hormonal." Better living through chemistry preceded better living through information. Hormones were the driving force in human behavior: they made us do things and especially made us feel certain ways. Some hormones are proteins, and proteins are involved in making other hormones, so the link is there, it's just pushed back one step.

How crucial is that one step back? If you listen to some geneticists, it is the ultimate step in understanding *life*. All life forms begin in DNA; all life therefore is DNA, and maybe all life *is*, is DNA. Know the DNA and you know everything. They—the geneticists who have pushed to translate the full length of DNA into the letters

ATCG so that it can be "read"—sound religious, awestruck, overwhelmed by the power and the majesty of the DNA: it's the Bible, the Holy Grail, the Book of Man. Those are their very words. Our fate, we are told by James Watson, one of the formulators of the double helix, lies not in our stars but in our genes.

On a more mundane, workaday level, the DNA is called a code, an encyclopedia, an instruction kit, a program, and most often, a blueprint. Well, sure: I started out as just a little cell with a nucleus of DNA, and here I am. Those must have been plans in there.

But it took about 50 years to get where I am now, and I think that is worth thinking about. When you have the code or the plans for a person, you don't have the person. If you take my DNA and clone it, make copies and start those copies growing into people, it will still take 50 years to get where I am now. And a lot has happened in those 50 years.

Plans and blueprints imply a building project. If I want to build a cabin, and I have plans, I could get a bunch of people and machines and do it in a few days, or I could work with hand tools by myself on weekends for the next ten years. A blueprint doesn't have time built into it; a blueprint isn't about growth, and so it isn't a good analogy for DNA.

A recipe might make more sense as an analogy. Take bread baking, which combines making something with growth, the growth of the yeast that gives bread its rise. The same recipe under different circumstances gives you different breads. Use a flour from a wheat grown in one part of the country and you have a different mineral composition than that from flour grown somewhere else. Bake on a humid day and you get a heavier bread than you would on a dry day. Bake on a hot day and it rises faster and has bigger air holes. Bake the same recipe every day for a week, and no two loaves will be exactly the same: the web, that distinctive pattern of holes, will vary from loaf to loaf. Bake it in different pans or in different ovens and you'll have differently textured crusts.

Why is it that I think of bread baking as a way of understanding DNA and growth and development, and most of what I have been

reading refers to blueprints, plans and information? Gender? I'm not
arguing that bread baking genes lie on the X chromosome or blueprints on the Y, but there is an experiential element here, maybe a
culture of gender. I've never worked with blueprints—actually, I
doubt that most of the men who've used the blueprint analogy have
either. But I have baked a lot of bread in my life. I've seen, over and
over again, how making such a comparatively simple thing is so complex, so varied, so nuanced.

I started baking bread at a really lousy moment in my life. I was
angry, disgusted with my work, disappointed in myself. I turned to
the kitchen for solace, for another arena in which to accomplish
something. And I let the anger and rage go when I pounded that
bread. It was my husband who noticed when the bread wasn't as
good that I must be doing better, emotionally. Forget baking with
love—I knead better when I bake with anger. I bake once in a while
now, sometimes with anger, sometimes with affection, sometimes for
holidays. And people comment on the breads—you've outdone
yourself this year, that challah is as smooth as silk; or, how lovely, but
heavier this time. I start with the same ingredients and the same
recipe, and I end up with something different each time.

I've also made a couple of babies in my life. And while I understand about the DNA and the plans and the blueprints and all that,
I also remember the rituals of avoidance. I read the ingredients on
every food package before I used it, hesitating to put things I couldn't
pronounce into the growing body of my baby. I avoided coffee, and
cat-litter dust, and by the time of the second pregnancy when the
rules had changed, alcohol. I read books about fetal development
until I decided it was totally incapacitating: how could I possibly
walk down the subway steps, or cross a street and breathe exhaust
fumes, or even get out of bed, on arm-bud-development day? All
that information management and blueprint-following wasn't happening off somewhere else—it was happening in my belly, day after
day after day after day.

This information, these genes, this blueprint—it is always located in a place and in a time. The same DNA, the same genes, will

make a tall person or a short one, depending on the nutrition available during growth. You could have all the symptoms of copper deficiency because of an error in the gene involved in using copper—or because you're in a copper-deficient environment. It was someone who noticed the brittle wool of sheep grazing on copper-deficient land who identified the genetic disease that involves the inability to metabolize copper and has brittle hair among its symptoms. The same DNA will produce sheep with lush wool in one place and short, unusable wool in another. The same DNA, the same genome, will produce a green grasshopper in my backyard and a locust somewhere else. Monday's bread is lighter than Tuesday's; Tuesday's crust is crisper than Wednesday's.

It's the element of time that seems so strangely absent in the discussion of DNA and genes—time and process. One of the Nobel laureates in genetics likes to hold up a compact disc at his lectures, and say, "this is you." As if those genes that were there 50 years ago at the moment of zygotic zero when I started, and the me that stands here now are one and the same. In the talk of the geneticists, time seems to come in only in discussions of evolution, of "progress." They do not include time in its daily, processual, experienced way, the time in which I bake a bread, grow a baby, the time in which some sheep uses the copper in the soil, the time in which an egg experiences the open spaces of my backyard and turns into a grasshopper or feels the crowding and becomes a locust.

If DNA is a bible, it's capable of being read in a lot of different ways. My friend Eileen Moran says to think of the DNA as a musical score, notes on a page, but capable of so many nuanced interpretations. Is it possible that this static thing, these notes on the page, this string of ATCGs, *is* life? Or is life the process itself in which these—and other—notes are played?

And what is a bread? Is it the recipe? the ingredients? the process? Or is it all of those, the end product of a complex interaction played out over time?

And what am I? Also the end product (or product-to-date) of complex interactions played out over time.

Prediction and Uncertainty

Think for a moment about identical twins. Identical is a funny word—it means they're the *same*. And so they are, strikingly so, which is why people noticed that they are a particular kind of twin. Identical twins grow from the same fertilized egg. One zygote becomes two people. Or, even more rarely, three. At the moment of zygotic zero, when the egg and sperm join, they are not only identical, they are one and the same. Very soon after that they split, and go their separate ways.

If the splitting occurs within the first three days, each develops its own placenta. If they split a bit later, between the fourth and the eighth day, they will, each within its own amniotic sac, share a placenta. Later still and they will share a single sac, too. And later than that and the twins themselves may be joined, "Siamese" twins, two people in a more-or-less shared body.

How identical are identical twins? Right there, made of the same fertilized egg, in the very same woman, nestled in the very same womb, they do go their separate ways, experiencing life differently. Genetic changes take place, as the eternal splitting and building of cells occurs. Mutations occur differently in each. One is born a boy, with the full XY chromosomes; the other, missing the Y chromosomes in many of its cells, is born a girl. A girl with one X and no Y chromosome, a condition known as Turner's syndrome, turned out to be the "identical" twin of her brother, grown of the same fertilized egg.

And even without dramatic mutations occuring, twins are not the same baby twice. At birth they do not weigh the same; their rates of growth can be dramatically different. Placed here or there, in the same womb in the same woman in the same environment, they're not having the same experience. *Here* isn't *there*, and nothing is ever the same.

They don't even have the same "genetic" diseases. Type I diabetes, for example, has been traced to a gene located on the small arm of the sixth chromosome. One variation on that gene, one allele, "causes" autoimmune diabetes. But genetic causes are almost always expressed in probabilities, in increased odds. In one identical twin the gene "causes" the diabetes. In the other, it doesn't. If one identical twin has type I diabetes, the chances of the other twin having it are only 30 percent. More than two-thirds of the time the identical twin will not have this "genetic" disease. Some genes are said to be more powerful, their effects more evident. They have what geneticists call higher "penetrance," and the odds of a shared disease or trait go up. Some are less powerful, their effects less evident. They have lower penetrance, and the odds go down. Diseases, traits, characteristics— one fertilized egg, two people, two individuals. You can't predict which will be which, how life will play itself out over time.

You can't get away from the idea of *chance.* Start with the same egg and the same sperm forming the same zygote in the same woman, and one twin gets sick and asks, "Why me?"

That is *not* what we used to think science was about. Physicists have had to come to terms with that uncertainty, raising it to a principle. Classical physics thought you could know it all, "that our cosmos is governed by mathematically precise laws at all scales, from the inside of an atom to the totality of the universe." And then chance raised its head, and "the positions and movements of the invisible atoms that make up all objects were now lost in a probabilistic blur. While the universe remained determinate, the revised mathematical constructs of physics—more accurately reflecting the way atoms behave—could no longer promise completely predictable events and objective, universal knowledge." You can't, it seems, know it all.

Over and over again I find myself drawn to the question, What is the point of mapping genes, of understanding genetics? What objective, universal knowledge is to be had in the reading of the genome? What predictable events are there? Every so often I see some person suffering because of the consequences of some gene gone awry, and I think, it would be a good thing in the world to fix that. I am reminded that good can come from all of this. And probably some good will.

But researching genetics and mapping the genome are only partly about the hope of fixing things, of making the sick well. They are even more about the drive to know, to understand. This research is very deeply about understanding and knowing how it works. That is why the scientists flirt with the language of spiritual quests, talking about "our fate," talking about the "Holy Grail," and the "book of life." To know, to understand, to unlock the secrets of the universe—that's what we've been after for a long time, looking in a lot of places. Burning bushes. Palm lines and tea leaves. Crystal balls. The tangled strands of DNA within, like the stars above or the atoms everywhere, are a good place to look.

But how do you *know*, how do you know you know? I watched a weather report with my kid. On a Monday they were predicting rain for Wednesday. "How do they know that?" she asked. "Can they feel it in their bones, or are they just guessing?" You know you are right when you can predict as well as explain, when your knowledge plays out in the world. Weather reports aren't really a bad comparison to genetic testing. It's not that they feel it in their bones, and it's not that they're just guessing, and it's also not that they can predict it all that well. There's a good chance of rain on Wednesday. Though maybe it'll get all clouded over and never actually rain on Wednesday. Or maybe Wednesday will be clear but it will rain on Tuesday. Or it will rain in Manhattan but not in Brooklyn. It's a prediction, based on knowledge, but also carrying with it the knowledge that you can't *know*.

Once we thought that the weather reports would get better and better. And maybe, with new information and new information-management techniques, they have. But we still have to make decisions about umbrellas based on inadequate information, and it seems more and more likely that we always will. There is too much that remains to the play of chance, to the inevitably unpredictable randomness of the world.

At first it looked as if unlocking the genetic code would tell us which diseases would strike which people. And it does, some of the time. It may get better as time goes on and more information is gathered and managed. But the inherent unpredictability is always there. Take one cell, one "blueprint," one "book" of DNA, and twin it or clone it as you will: you don't get the same person over and over again. Read all the information, decode the book, and you can make predictions, based on knowledge, but you also must acknowledge that you can't know. Decisions will have to be made—about screening and testing and aborting and preventing and treating—always with inadequate information.

Why, we want to know, why does one person get diabetes and another not? Why is one tone deaf and one musically gifted? Why am I this way and you that? *Why?* Children ask why all the time. It is a powerful, beautiful thing, that search for *why.* But you can't always satisfy it, and that too is a lesson to be learned. My great aunt Joan had an intellectual turn of mind. She was a nurse, a professional, and she had her children later in life. She was determined not to hush those endless "why's" children asked, but to answer, each time, to the fullest of her ability. To her sisters' amusement, the endless answers just led to more endless "why's." And so eventually Aunt Joan learned sometimes to answer, "Because the chicken told a lie."

Why did a hundred people die in a storm in Brazil? Because a butterfly flapped its wings in China. Why does one twin develop cancer or diabetes and the "identical" twin not? Because a cell split one way and not another. Why am I who I am? Because the chicken told a lie.

On Authority

It is important to understand how far from perfect the predictive nature of genetic information is. And it is just as important to understand how far genetics is from control: in that way too, genetics is a lot like the weather reports.

But it is also important to recognize that genetic information, made available to us through genetic screening and testing, does offer some information, some predictability. Sometimes the prediction does approach the absolute: some versions of genes are lethal. If a baby inherits one of these forms of a gene, it simply won't survive. Most of the time the predictions are far more erratic, best expressed as probabilities: a 30 percent chance of type I diabetes; a 50 percent lifetime chance of ovarian cancer.

Genetic screening or testing starts with the gene, the genotype, and then makes predictions, however vague or specific, about the phenotype. These predictions—inaccurate, incomplete, uncertain as they are—open up the enormous quandaries of contemporary bioethics. But we also use genetics the other way around: this outcome, this phenotype, implies something about the genotype. This is most usually called "genetic determinism," saying that something—a physical trait, a characteristic, a skill, a disability—is caused by genes.

Genetics, as we use it, is a way of understanding life, "human nature," behavior, being in the world. But genetics is also a way of shutting our eyes to understanding. Genetics is also a way of avoiding explanations.

Saying "it's genetic" is like throwing your hands up in the air; it's a fatalistic, deterministic "explanation" that is no explanation at all.

When my oldest kid, Daniel, was in first grade, he came home all excited about a science experiment that they had done in school. He very carefully set it up for us in the kitchen, measuring baking soda and vinegar and creating quite a little flurry of activity in a glass jar. And when my husband and I admired all of this, and asked him, "So Danny, how does that work?" he answered, shrugging his shoulders, "I don't know—it's science!"

That is science-as-magic. Most of the time that kind of story presents a lesson about the necessity of understanding science better. We are continually told that if we understood science better, we wouldn't have foolish fears and superstitions. We wouldn't fear crazy science fiction scenarios because we would understand that science is no danger to us. For all of my working life in the sociology of biomedical technology, what I've heard continually from the scientists and the doctors and the technicians is that to know them is to love them: if only you understood what we are doing, you would appreciate our work.

But genetics itself has a science-as-magic quality. For all practical purposes the study of genetics often works like a surprisingly accurate and very sophisticated tea-leaf reading. One of my favorite cartoons shows a turbanned "Madame Rosa" replacing the "Fortune Teller" sign in her front window with one that says "Geneticist." For many of the "gene for" something-or-others that we've been hearing and reading so much about, they haven't even found the gene itself. They've found a marker, an indication that somewhere in that neighborhood on the chromosome there must be such a gene because people who share that marker share some illness, condition or phenomenon. It's a prediction, and it might work fairly well in some cases, but it is in no way an explanation.

In science as in life, prediction does not actually depend on explanation. Danny could predict a pretty good show in the jar for us, without any kind of explanation of why vinegar and baking soda would do that. All the livelong day we rely on predictions that we can't always explain. Part of that is because of the role of technology in our lives: nobody has the time or the energy to understand all of it.

Some of it you just use, predicting that the alarm clock will work, the freezer will freeze the ice cubes yet again, the plane will stay up in the air. That is the appeal of those "how things work" books—we take an awful lot for granted in the world of technology. But there are other kinds of predictions we make, predictions that we cannot necessarily explain. Something happens that I know my sister will find funny and my mother will not. Can I explain that? Not in any meaningful way, just by saying it happened that way before and so it will happen that way again. It's the "kind of thing" my sister finds funny and my mother not. Why? What in their history, psyche, (genes?)?

Because the technology involved in the tea-leaf reading of genetics is so incredibly complicated, the essentially primitive nature of the predictions may elude us. This struck me forcefully some years back when I visited a lab where prenatal diagnosis was done. They worked with aspirated amniotic fluid, the water in the womb suctioned out with a needle. Cells from the fetus float in that water. Those cells were cultured, and then magnified and photographed. The next part of the process was cutting out the individual chromosomes and pasting them in size order on a piece of paper, counting how many pairs there were and seeing if there were any extras. Literally—cutting and pasting and counting. If there were three 21st chromosomes the fetus could be diagnosed as having Down syndrome. Three 18th chromosomes and the fetus had, by definition, the condition known as trisomy 18.

I'm not belittling how complicated it is to get to the cut-and-paste portion of the diagnosis. Suffice it to say that up until the mid-1950s geneticists thought there were 48 rather than 46 chromomes. So it is a complicated technological feat to photograph, enlarge and count those chromosomes. But it is not an explanation. To diagnose, to predict Down syndrome is not to explain Down syndrome. And to predict the fact of Down syndrome is not to predict the experience of Down syndrome. One fetus so diagnosed is not strong enough to survive the pregnancy; another is born with grave physical and mental handicaps and dies very young; and another grows up well and strong and stars in a television show.

Even where genetics does offer understanding and explanation, that doesn't necessarily improve the predictions, a point I made earlier and will keep coming back to. Sickle-cell anemia is understood thoroughly, from the single misspelled base pair in the DNA through to the misshapen red blood cells clogging the capillaries, and still they cannot predict which person will be severely and which mildly affected by the disease.

So what genetics has to offer us is prediction, with greater or lesser accuracy and greater or lesser specificity. Prediction is a concept that you and I are familiar with. We can have ideas about predictions—legitimate ideas, backed by solid values we hold firmly and dearly—that are not dependent on the particular technology in use. Judgments about the worth, value and danger of genetic predictions do not require an understanding of the science underlying the prediction. Moral authority does not rest on technical authority.

A person might claim that this is not a good historical moment to predict the future sexual orientation of a male fetus. And that claim, that belief, that value judgment is not dependent on the technology. Tell me you can do it with a brain scan in utero, with fetal cells withdrawn from amniotic fluid, or by playing Judy Garland tapes at the pregnant belly. One's judgment about whether or not this is a good kind of prediction to be making doesn't depend on the technology that does the predicting, and it certainly doesn't depend on one's ability to understand the technology.

This seems so obvious, so clear to me. And yet we permit ourselves, over and over again, to be intimidated by the technology, by the science of it all. I've attended countless conferences on genetics, biotechnology, ethical issues in genetics, and the like. And virtually every one of them starts with a scientist explaining the technology. The lights go down, the slides go up on the screen, and the technological razzle-dazzle begins. Then, and only then, is the discussion opened up to "ethical" or "social" concerns.

The technical presentation is supposed to open up the discussion, but what it does is silence people. People want to talk about the moral and ethical concerns they have, they want to talk about the consequences of these new technologies in their lives and the lives of their children. They want to talk about how this work might change the world, for better or for worse, and how we might control it. They want to talk about what the technology means in their lives, and instead they're told they have to understand the technology before they can judge it.

My family spends a couple of weeks each year in a cabin in the Adirondack Mountains, an interesting mix of primitive and modern. It's rather like a wooden tent—has a plank floor and wood walls and a roof that keeps out rain and a screened-over opening to the outside with a wood flap to close when it gets too cold. There's a pipe bringing water to a faucet outside the cabin. There's a two-burner propane stove inside, and a fire pit of stones outside. A refrigerator has been wired in but there is no other electricity—so you have the advantage of the storage and don't have to keep driving into town for ice. There are kerosene lanterns and we bring flashlights. Toilets and showers are up the hill.

When we're there we allow ourselves to be moved into a different kind of rhythm. We have leisurely breakfasts, spend long evenings playing games by kerosene lantern-light. The people who run this place have been encouraged to put electric lights in all the cabins, to save on their insurance costs. There's been a big debate raging at the beach by the lake for years now. Some of us say that's nice, we will be able to see better at night. And some of us feel something is being lost. We know we don't have to turn on the lights, and can still use the lanterns, but . . . Somehow you don't, or it doesn't feel the same, or something.

I enter into this debate, and having been there every year for over 20 years, feel qualified to have an opinion. I don't, when you come right down to it, actually understand how electricity works. I hesitate to come right out and say that. But after all of my own science courses, after god knows how many science fairs, projects and homeworks with my kids, I don't think I could build even the most primitive of generators without getting out a book and looking it up all over again. Let alone wire up the Adirondack Mountains. And I don't, I'm embarassed to admit, actually know much about kerosene—do they manufacture that? Distill it? Out of what exactly? And matches—wood, sulfur and what?

And yet, in spite of all of that ignorance, I know what the consequences of electrifying the cabin might be and I know what the consequences of using kerosene lamps have been. Make a fool of me if you want to with a bunch of technical questions, but don't tell me I don't know what I am talking about.

I don't tell you this story in praise of ignorance, neither in matters electrical nor in matters genetic. Rather I tell it to help place technical knowledge in the appropriate context. There are relevant technical questions to be asked and answered in the decision about wiring the cabin: the pollution caused by the kerosene compared to that caused by electricity; the actual fire risk; and so on. But if we on this hillside decided to have a big meeting on the beach to discuss it, should the meeting start with a presentation by an electrical engineer on how electricity works and how he would go about wiring our cabins? I'm sure he could tell us how it is done, and how safe and clean and cheap it is. And that would shape the discussion that followed.

When I go to a genetics conference to discuss the bioethical issues involved in, for example, finding the "gay gene," and the meeting starts with the geneticist explaining how he did his research, how he ran those gels, how the science and technology of it all works—that too shapes the discussion that follows. We don't discuss "Should you do this?"; we discuss "How did you do this?" We don't discuss "Do we want to know if there is a way of predicting sexual orientation?"; we discuss how well we can predict sexual orientation.

Is research into human genetics fundamentally a social, political and ethical concern, with some technological obstacles? Or is it fundamentally a technological concern, with some social, political and ethical obstacles? The official American version has been the latter: three billion dollars of funding goes to the technical work, and 3 percent is tithed for "ethical, legal and social implications (ELSI)." Arthur Caplan, a prominent bioethicist, has called ELSI the full-employment act for bioethicists. It came about as the response of James Watson, co-describer of the double helix and first director of the Human Genome Project, to the ethical questions being raised at a press conference. ELSI, the cash cow for ethics, stands in an inherently contradictory position: a statement that those concerns are being taken seriously, and a statement that they are not to be taken seriously enough to interfere with the project.

Bioethics generally, and ELSI specifically, ends up in the middle, between biomedical science and research on the one hand, and the concerns of the public on the other. But with all the power and the big money in the hands of the science, bioethics becomes a translator, sometimes an apologist, sometimes an enabler, of scientific "progress."

At bioethics committees and conferences and discussions people start off talking about moral wrongs and end up talking about regulatory oversight. They start off talking about playing God and end up talking about extra filters in the laboratory safety hoods. I never quite understood the process until John H. Evans, a graduate student in sociology at Princeton, introduced me to the idea of "moral Esperanto" in the bioethical discussions.

Esperanto was conceived as a shared language that a multitude of ethnic and linguistic groups would develop and use in the market. Like bioethics, it would be its own world with its own language, a neutral place where competing moral and ethical considerations could play out. But Evans shows that Esperanto is a language learned from the colonizers, and that in framing the problems, in expressing them in a particular language, certain solutions come to be seen as inevitable.

The example Evans uses is the concern theologians raised about the prospects of human genetic engineering. They worried about "playing God," about people taking control over parts of life that people have no business controlling. They worried about doing wrong, about violating moral principles. Their concerns were raised in a joint letter to President Carter, and a presidential commission was established. Evans has done an analysis of that process, from the first letter through the committee working on five drafts, to the production of the final report, "Splicing Life." What Evans shows is the way the process of translation into a "common" tongue ends up leaving people speechless. At each point the theologians' concerns were translated, and lost everything in the translation. The commission objected to "vague" concepts, which meant just about everything the theologians might say. If your concerns are on the level of principles, values and morals, then they're vague. If you're worried about safety tolerances, you can be specific. To eliminate vagueness, "playing God" is translated as acting without knowing the consequences, taking risks. So controlling the risks becomes the solution. Bacteria are being genetically engineered and that creation of a new life form is playing God? Playing God is taking risks? Put more filters on the lab hoods and then you're not playing God? Somehow, that was not what the theologians intended.

Speculations weren't to be permitted, either: the theologians were worried about something that hadn't happened yet. Their greatest concern was about germ line therapy—genetic therapy that changes the cells that the treated individual will pass on to future generations. That was a central concern: scientists would do something to a person that would change future evolution. We would be, in a new, direct, intentional way, controlling our own evolution. The commission decided that since it hadn't happened yet, they couldn't evaluate germ line therapy, they couldn't judge the circumstances in the future in which people might want germ line therapy. That of course was precisely what worried the theologians: this was something they thought was wrong, and worried that it would come to pass. The commission said it couldn't address it until it had come to pass, that you cannot make judgments on speculation.

I've seen this technique used dozens of times: someone in the audience or on one of the panels at an open discussion of bioethical concerns raises some concern over something like human cloning, or parents doing selective abortion for gay fetuses. The scientists quickly speak up: that isn't possible, they reassure us, you don't understand the genetics involved. Five years later, of course, that *is* possible, and then it is too late to decide whether or not to do it: we wake up to find it done.

The theologians, whether or not you and I agree with them, are making a deontological argument: some things, they say, are wrong. Period. End of sentence. Germ line engineering, gene therapy that changes the gametes of the treated person, is wrong. The commission translated that, worked with that, massaged that, spun that. What came out was that people lack the knowledge of possible outcomes of genetic engineering, and thus genetic technology is not claimed to be wrong as such, but wrong because of its potential consequences. From the deontological position of absolute wrong, the issue had switched to the knowability of the consequences. As Evans summed it up, "God had become a consequentialist."

Bioethics has become its own kind of technical language, its own form of mystification. Mystification is a political tool: making something complicated is a way of disempowering people. I'm a sociology professor; I get paid to read. I can afford to take a couple of years and read in genetics and bioethics. Most people probably cannot do that; they have other things to do. But the conclusion that I have come to, from all of that technical reading in genetics and in bioethics, is that you don't need the technical understanding to make the moral judgments.

A group of sociologists in Scotland came to the same conclusion. They ran focus groups of lay people on ethical issues in genetics. They concluded, "Technical competence was neither relevant or important to the majority of participants in our study: they discussed issues without need to display technical competence. When technical issues were mentioned, the accuracy of the knowledge was irrelevant to the point being made." They gave an example of a group discussion in a working-class area of Edinburgh: "They are going a little far. If they want to go an' investigate the DNA system and find oot that OK somebody's gay because there is a little slip-up in the XY hormone, we can do an injection and fix that, or a kid's going to be born mongoloid, rather than abort we may be able to find a way that we can actually sort the gene oot. We are getting to the part with genetic engineering if somebody is going to get a deformed child then they just get rid of it and say 'right the next one that you produce will be.' "

This person is completely wrong on every technical point going. XY isn't a hormone; mongoloid isn't the current word and it's not a "gene" to be "sorted out." And so what? The question that the person is raising is about drawing moral lines, about drawing lines and going too far. Again, you or I may or may not agree with him, just as we may or may not agree with the far more sophisticated language the theologians used. But moral authority does not rest on technical authority: the concerns that are being raised, including the concerns that you personally may feel, are in and of themselves worth discussing.

Genetics, as a science, as a practice and as an ideology, is offering us a great deal. But we have to decide if we want what it has to offer. Those decisions are not technical matters. The technology of it all *is* overwhelming. Keep bandying about terms like "alleles," "RFLPS," "clines," "22Qlocus," and most of us are left in the dust. Promise a cure for cancer, an end to human suffering, and it's hard to argue. As sociologist Troy Duster puts it, "Technical complexities of vanguard research in molecular biology and the promises of success incline us to go limp before such scientific know-how."

We cannot afford to go limp. We'll be carried off to places we might very well choose not to go.

Maps are being drawn, maps of the human genome, and with every map comes a claim to territory. I see three arenas in which genetics proceeds: evolutionary history; illness; and "genetic engineering," or controlling the future. For each of these concerns we need more than maps: we need an imagination. It's the kind of imagination C. Wright Mills wrote about, an imagination that can see the connections between our own personal troubles, our daily lives, our intimate concerns, and the world in which we live, the issues that face us collectively. We live a biography, a personal tale, but we live it in a moment of history, in a collective time, and it takes a leap of imagination to understand the connection.

The first area is evolutionary history, the story of how we came to be, the collective origins tale we can tell. Genetics promises to read us evolution, to map human history. In one way it is the promise of the new genetics that comes with the fewest strings attached: history can't be patented, evolution can't be marketed. There is no profit to be made. But in another way, it comes with the most strings: it comes in the tangled packaging of race. Race has been the only way we have had, for a long time, of talking about the differences and the relationships between the many and varied people of our world. And race was invented out of another moment of historical research and scientific exploration. Race is not, as I have often been reminded while working on this project, a system of classification; it is a system of oppression. There has never been, and I can't imagine how there ever could be, a way of classifying the peoples of the world that isn't also a way of controlling people. As the science and the technology get more impressive, the historical burdens of race get no less complex, no less impressive. We have no way to confront our history without standing in our history, no way to look back but from where we now stand. As the white mother of two white children and one black child, I stand in a particular place, as each of us does, and exercise my own imagination placed in my own map, and invite you to explore with me the implications of the new genetics for thinking about race.

The second area of work in the new genetics is illness: genetics promises to revolutionize medicine. Whether it can or cannot accomplish that, genetics is revolutionizing the way we think about the body. The solidity of the body is breaking down as we imagine activity at the cellular level. The wholeness of the self is fragmenting as we think about lists of instructions, the pages and pages of ATCGs that make up each gene and the three billion such letters that, we are told, make up each of us. Sickness defines wellness, as black defines white, as all things take their meanings from what we define as their opposites. It is cancer, when the instructions go "haywire," that increasingly defines normal for us, the body as it is supposed to be, taking its meanings from the body gone astray. And I stand here, amidst the cancer losses of my own life, in my middle-aged body, looking at the as-yet unscathed bodies of my children, and exercise my imagination as I explore this new map of the body that genetics is writing.

But most profoundly what the new genetics is offering us is what the clergy called "playing God." It offers us the possibility, the hope and the fear, that we can control the future. Genetics works as tea-leaf readings for the future, offering us predictions and, only sometimes, choices. We're being asked to think about what kinds of children we want to bring into the world, what kinds of people we want on our planet. We're being asked to look at all the bits and pieces of ourselves, and choose which ones to keep, which to discard in the generations to follow. And I stand here, on the brink of this future, and know there are no maps. There is only imagination.

Part I

MAPPING THE PAST: THE MACROEUGENICS OF RACE

Inventing It

Race is, and has always been, about difference, about otherness, about othering. The myth of a "first encounter" is so powerful that it is hard not to think in terms of a moment of meeting, where two races faced each other. But there was not a moment of encounter: there have been repeated moments of encounter in a dance of coming together and moving apart that is the history of humans. We didn't spring up as separate, pure, unsullied races in separate places, only to come together in the great American melting pot. In the greater scheme of things, we were all cooking in each other's pots just yesterday. Mixing is the rule rather than the exception in human history.

But people don't live in the greater scheme of things, they live in the here and now, products not only of a biological history, but of a cultural history. And the image of the British guy pulling up on the foreign shore, the Dutchman sailing into the faraway port, the ship looming out of the sea and bringing the ghostly white invaders— these images are part of our here and now. While in the larger evolutionary time the various people may have been one but yesterday, those moments of encounter are firmly fixed in cultural memory. And while surely there was recognition, as pale eyes and dark eyes met, as one human face searched another, there was also the sense of strangeness. What history makes clear is that in the encounters between the lighter and the darker peoples of the earth, the lighter did not behave well. And what we are left with is not a story of an encounter, but a story of colonialization, power and domination.

The invention of race was a response to encounter, even as it was a weapon used in domination. Jonathan Marks, who is both a biologist and an anthropologist, in his excellent book *Human Biodiversity: Genes, Race and History*, credits the eighteenth-century botanist Carolus Linnaeus with the first "scientific" classification of humans into large, distinct groups, what we now usually call "races." For Linnaeus, the classification of humans was part of a larger project, the organization of all life forms. Linnaeus's contribution to biology was to move away from the "Great Chain of Being," the ranking of all species on Earth from the lowest to the highest, and to substitute the clustering of species we now recognize. For example, some animals, he noted, share the ability to nurse their young: he designated them the class Mammalia, distinguishing them from other classes of animals which do not nurse. Within this as within all classes, he distinguished "orders." "Ultimately, therefore, every species was a member of a genus, a member of an order, and a member of a class."

And so too Linnaeus classified the varieties of humans as "similar in kind if not in magnititude to the zoological genera in an order, or the zoological orders in a class." But classifying people is not and never will be, as Marks reminds us, like classifying clams. Linnaeus's goal was to establish what the natural categories of the human species were. He found five natural categories of humans, four of which were geographically based (in the Americas, Europe, Asia and Africa). With each category came a description of the essential characteristics that define the people. *Homo sapiens Americanus* was "red, ill-tempered, subjugated. Hair black, straight, thick; Nostrils wide; Face harsh; Beard scanty. Obstinate, contented, free. Paints himself with red lines. Ruled by custom." *Homo sapiens Europaeus* was "white, serious, strong. Hair blond, flowing. Eyes blue. Active, very smart, inventive. Covered by tight clothing. Ruled by laws." *Homo sapiens Asiaticus* was "yellow, melancholy, greedy. Hair black. Eyes dark. Severe, haughty, desirous. Covered by loose garments. Ruled by opinion." And last (and obviously least) *Homo sapiens Afer*: "black, impassive, lazy. Hair kinked. Skin silky. Nose flat. Lips thick. Women with genital flap; breasts large. Crafty, slow, foolish. Annoints himself with grease. Ruled by caprice."

If that sounds like a 1940s casting call for a B movie, the fifth grouping belongs with the 1950s science fiction genre: coneheads and flatheads, the large and the small, the remote, the curious and the imaginary: *Homo sapiens Monstrosus,* scattered from China to Canada and beyond.

These, then, are the basic groups: red, yellow, black, white and very strange.

The categories shifted around some over the next two hundred years, sometimes subdividing into more and more minute race categories—in anthropology, Carleton Coon had ten white races by 1939—but the fundamental question remained the same: What are the natural, essential races of people? The question dominated physical anthropology from the 1700s until the mid-1960s, a question that Marks calls "anti-anthropological, anti-biological and antihistorical. In searching for the divisions of the human species as a cardinally biological question, the question assumed, and in turn legitimized the proposition that the human species could actually *be* divided into a small number of basic biological groups."

More sophisticated groupings separated out, or tried to separate out, the biological, physical variations (skin color, hair texture, lip shape) from the cultural variations (personality, social organization). With fancier technology, you could get elaborate measurements of things like skull size and shape, bone length, dental variations and blood-group frequencies. Nazi science was fond of various kinds of skull- and nose-shape measures to identify Jewish racial characteristics. Straight through the growth of the eugenics movement, and out the other side of the Holocaust, physical anthropology searched for the essential groupings of humans.

Which people belonged in which racial groupings? How many racial groupings are there? What are their origins? Those were the questions. What remained unquestioned was the category of race itself.

Seeing It

Race is one area in American society in which the discussion is already well grounded in moral rather than in technical authority—unlike in so many other areas of genetics. Illness, disease, making babies, and the body itself have been mystified and medicalized, and basic concerns about life and death have been ceded to technical authority, technical competence. Doctors sign our birth certificates, and eventually pronounce us dead, ushering most Americans in and out of life. In America today we think we need someone to tell us if we're sick or well, pregnant or infertile, healthy or dying. Cancers, illness, procreation, infertility, genetic disease— all are terribly complicated, Americans think, and doctors are one of the few authorities left whose orders most Americans are willing to follow—or think they should follow.

But race? That we think we understand. The biologists and the anthropologists now tell us that there is no such thing as race, that racial categories do not exist for human beings. But we Americans nonetheless recognize as white the white woman behind the counter, as black the black woman next to her, and as Asian the Asian customer approaching them. We see race and we understand it as a "genetic" characteristic. We have disagreements as to what race does and ought to mean, but we have a remarkable consensus on what it is, without any ability to define it technically.

Even young children seem to understand race, to recognize racial differences. In one of the sillier arguments that everything is genetic, one book argues that while race itself is not a useful biologic category, the ability to see it is. Using data that show that children as young as four see race as more likely to be inherited and to remain unchanged over the life span than either occupation or body build, Lawrence Hirschfield argues that there must be a pre-existing, genetically based specialized cognitive facility for racial thinking. That is, four-year-old kids have noticed (not all of them, and not consistently, but they are taking note of this at more than chance levels) that black people are more likely to have been black children themselves, and to be the parents of black children, than fat people are to have been fat or to have fat kids, and more likely than nurses are to have worn nurses' uniforms as children or to parent children who wear nurses' uniforms. And white people too tend to have once been themselves, and to be the parents of, white children. This also four-year-old children have noticed. But since they couldn't possibly have had an opportunity to learn that in their four short years of being here, Hirschfield argues, there must be a gene for noticing it, a genetically based domain for recognizing racial categories.

It is silly, of course it is. Four years is a long, long time. It is remarkable what children do, and do not, learn in four years. When my oldest kid, Danny, was about that age, maybe a bit older, he was friends with Benjamin, a little blonde kid down the block. One day Danny and Benjamin went to Benjamin's house to ask if Benjamin could stay for dinner. Danny came back alone, saying Benjamin's grandmother said no. I knew that household—there was no grandmother staying there. But Benjamin's mother was heavy, and I know that kids sometimes confuse heavier, larger people with older people. "Is Benjamin's grandmother tall, like me?" Yup. "Is she kind of fatter than I am?" "Does she have curly black hair?" Yup. Danny, I explained, that wasn't Benjamin's grandmother. It was his mother. No, Danny told me. It couldn't be. She was black.

Children understand the heritability of race—white, blonde Benjamin couldn't have a black mother, no. A black grandmother, yes.

In my household now my black child has a white, blonde grandmother, and we make no further assumptions about anybody's race or family. And that child, at the age of two, accustomed to suckling on my white breasts and knowing a lot of black people but never having seen them undressed, saw a black mother nursing her black baby. "Look!" Victoria shouted, all excited, pointing at the exposed breast, " 's brown!" She hadn't yet learned the consistency of race across the body, let alone across the life span.

What is it that the child, and the adult, are seeing when they learn to see race, this inherited, immutable, unchanging, fixed category of humankind? What characteristics are they picking out, and what meaning do they give to them? Hirschfield used the one most Americans think of as most basic: skin color. But race isn't always about color. The Nazis understood the Jews to be a race—not a religion, not a group of people united by how they chose to worship, but a racial grouping to be distinguished from the German race, the Aryans.

American ideas about race are always embedded in our history of slavery, in which race determined personhood. People of African descent were not people, not fully human, recognized legally as, at best, three-fourths of a person before the law, but without rights of citizenship, of legal and moral personhood. Race was carried in the blood: one drop of black blood marked a person, tainted a person, made a person less than a person, reduced a person to the status of an American black.

Other cultures have had other ways of marking differences between groups of people that make some people less than fully human, make one group dominant over the other. The British have long done it with class: a system of subordination also based on "blood," on "breeding." The elite, the people of privilege, could glide naked from the bath past a servant as if that servant were the cat. It was the servant's responsibility to turn away, not the elites' to cover themselves in the servant's presence. The servant *had* no presence.

We grow up in systems of domination, in systems of classification that serve for domination, and we take their codes for granted. We see race, and we see it as an obvious physical reality written on the body, marking membership in the community. We'd like to think that we see that difference, and *then* decide whether to accord it much meaning. In other words, we'd like to think that race exists, but that racism doesn't have to. But it doesn't work that way: to see race is to construct the system in which it exists; to construct that system is to learn to see race.

Seeing race is always about discriminating: a discriminating, discerning, trained eye recognizing the "essential" or defining characteristic(s) in the individual that confer(s) race categorization. And as with most things we've trained ourselves to do, it feels natural, the categories seem obvious. It is only when we confront an alternative categorization, another way of doing it, that the social nature, the constructed nature, the unnatural nature of the category is laid bare.

One such moment came for me a couple of years ago. I was going to spend six months in the north of the Netherlands, on a Fulbright fellowship to a university there. I was bringing with me my two daughters, then aged 13 and 5. It was the five-year-old, Victoria, that I was concerned about. She is black, African American, mine by adoption. We had until then lived only in Brooklyn—as racially mixed a place as one is likely to find on the planet. Would there be other black kids in her class, I worried. Would she be the only black kid around?

I was assured, repeatedly and sincerely by people I trust, that that definitely would not be the case. It would not be the case in any Dutch school; it would not be the case particularly in the international school where she would be.

She was the only black kid in her class. She was the only black kid I saw anywhere in that school.

If I hadn't been reassured by people I genuinely like and trust, I'd have just been angry. As it was, I was puzzled. I walked over to a wall of photographs of the school going back for years and years, group after group of class photos. No black kids. I didn't say anything, just kept watching, thinking about it. A few days later, light dawned for me: there were dark-skinned kids from India and Pakistan in all the classes. Black kids. European-style black kids.

For an American, with an American sensibility of race, Indian and African kids are not both "black." For a Dutch person, with a different race system in his head, these were all black kids.

So what does that story prove, anyway? That the Dutch draw a different line? Maybe between the Dutch and everyone else? Not being Dutch, are all the blacks, well, black? The Indian kids in her class could see what my kid and I could see, the distinctiveness of African features over and above the similarity of skin color.

So does the story tell us that race is a socially constructed category, constructed differently in different places? Or does it tell us that the Dutch draw their lines so tightly around themselves that they don't bother to make finer discriminations—not that they don't see or experience the distinction as existing, but that they don't see why it should matter.

And is that what white Americans do when they see a black kid whose family has been in the United States since slavery days, a black kid whose family arrived two generations ago from Haiti, and a black kid who just immigrated here from Nigeria, and calls them all "African American," seeing no meaningful differences?

But are those differences between these kids of African descent *race* differences? Can you see, see in the body and the features, *see*—not hear it in the accent, but see in the physical presence—the differences between these groups? If race is the biological, physical category we have claimed it to be, then we should be able to see (or as some have claimed, even smell) race differences. And if these are physical, biological categories, then you should be able to take away cultural differences—differences of language, dress, socialization and diet—and *still* see the differences between races. Because Americans quickly see the difference between, say, equally dark-skinned African and Indian people, Americans have called that a racial difference. But what about the differences between the Nigerian, Haitian and American blacks? Can you see those differences?

A Haitian can recognize Haitian people. Living in a neighborhood with a lot of Haitian people one can recognize Haitians. There is a "family resemblance," a recognizable, characteristic Haitian face. Am I treading on racist ground here?

So let me ask, Can you recognize an *Irish* person? And is that less threatening?

Years ago I told my older daughter, Leah, back when she was about five or so, that my friend Janet was coming by. "Which one is Janet," she asked, "the one that looks like Eileen?" I was taken aback. Does Janet look like Eileen? Eileen's another friend of mine—I'd never noticed any resemblance. And then I thought about it. Janet looks Irish. Eileen looks Irish. They were probably the two most Irish-looking people that had been in our house that year. Yes, I guess Janet and Eileen look alike. I hadn't made the connection because I took the Irishness of their faces for granted, made it the background against which their uniqueness stood.

And there we have the "they all look alike" business. And this too is a true story: Right after arriving in the Netherlands for our six-month stay, we met a neighbor who gave us a hand settling in, finding our way around the supermarket, and so on. She introduced us to her kids, including her son Jos, who could do some baby-sitting for us. Three or more times that first week, walking the streets of our neighborhood, exploring the playgrounds, finding the way to the school, Victoria thought she saw Jos ride by on his bike. She'd wave, call out—only to realize it was the wrong kid, the wrong blonde boy on a bike. And then, after one time too many, she turned to me in exasperation and exclaimed the inevitable: "They all look alike!"

Of course they don't all look alike—once you recognize that they do all look Dutch. Or Irish. Or Haitian. When you've learned to both see and not see the resemblance, then you see the uniqueness.

And now one more story from my globe-trotting experiences. I went to New Zealand, to speak to a midwives' conference there. The people who were in New Zealand when the white settlers arrived were the Maori. In the drawings and paintings reproduced in the guidebooks I bought, the Maori were a dark-skinned people. In the more recent photos in those same guidebooks, they were more or less my color, a darker-skinned white person's color, the kind of color that "tans well."

There was a special welcoming service on the day of the big mid-wives' conference. In Maori tradition, the members of the local *marae,* the traditional village of the area, welcome visitors to their land. It is a lovely ceremony, with women's voices singing back and forth, calling the visitors on to the home place, a welcome-and-response singing between women. The visitors, including me, walked slowly forward. The welcomers, the representatives of the *marae* and the midwives of the local area, walked slowly backward, welcoming us in. Then, seated in an auditorium space, the men took over, and men's voices moved back and forth. The man with us, the visitors, was a Maori accompanying the midwives. The men welcoming us were the men of the local *marae*. The entire ceremony was conducted in the Maori language—I just watched, enjoying it as theatre, understanding very little. Chief among the welcomers, next to a dark Maori man, was a British guy. A really, definitely British guy. He looked more British than Prince Charles—the ears, the color, the sandy hair. His Maori sounded smooth, without hesitations. Must be from the university, I figured, a local host representing the university whose campus we were on, participating in the ceremony in that capacity.

Turns out the dark man next to him was his uncle. I chatted with the sandy-haired man afterward, over cups of tea. He is a Maori who has to keep taking time out from work to show up at these conferences and things at the university all the time, to perform the traditional ceremonies, representing his *marae.*

I was beginning to catch on. These New Zealanders are a very mixed people: there's been a lot of mixing, and you can't tell who's Maori just by looking. And when I shared that insight with a local non-Maori woman, *she* was confused. You mean I couldn't tell that man was Maori? But his *nose,* surely I could see by his nose that he was Maori!

And so: seeing race is always about discriminating, a discerning, trained eye recognizing the "essential" or defining characteristic in the individual that confers racial categorization. From the enormous, vast confusion of features, shades, shapes, angles and planes, textures and colors that make up a human face, we construct categories. This configuration is Irish, this Maori, this Dutch. The configurations can be geographically narrow (Irish, Dutch, Haitian) or vast (African, "black," Asian).

When the configurations are narrow, they begin to approximate what sociologists have called "ethnic groups." A standard and classic definition of ethnic group can be found in a dictionary of sociology first published in 1969: "A group with a common cultural tradition and a sense of identity which exists as a subgroup of a larger society." This was written in an era in which "ethnic group" was a fairly unproblematic category: ethnic warfare was not constant headline material. The definition went on to say, "Probably most important is their feeling of identification as a traditionally distinct group." What my first sociology teacher and mentor Setsuko Matsunaga Nishi reminds me of is that ethnic groups are *vulnerable,* and that vulnerability—to "discrimination"—is what makes some ethnic groups into "minority" groups, in turn vulnerable to potential "racializing," being defined as a separate racial group.

One of the intriguing parts of the definition is its context in a "larger society." When ethnic groups feel themselves distinct enough, they may want societal control, separate nationhood, and ethnicity breaks out into nationalism. When they don't feel distinct at all, the identity begins to fade, the distinctions fall, and the "melting pot" takes over.

Ethnicity, Americans have found, is tenacious: it lives on in foods and rituals, in worship and holidays, in life passages and in ethnic community associations. The very history of discrimination may itself become the basis for ethnic unity. And to the extent that ethnicity leads people to marry and to procreate more within than outside of ethnic groups, it lives on in the body: it is possible to "look" Greek or Irish or Italian or Korean in America.

The sociologists' definition of ethnicity has tended to ignore that phenomenon, subsuming the "look alike" aspect of ethnicity under the category of race: "Ethnic groups should not be confused with racial groups. It is possible for an ethnic group to be a racial group as well, but often this is not the case." Race, Americans believe, is supposed to refer to biology, as if without culture; ethnicity is supposed to refer to culture, as if without biology. But race is a cultural, social category; and ethnicity is community writ on the body. I once led a group of sociology graduate students in a study group on adoption. One of the students, herself adopted, talked about not knowing anything about her birth family: "I don't even know my biological ethnicity," she said. For a moment we talked about whether she "looked" Italian or not, what her biological ethnicity might be. And then someone said, "Listen to us! Biological ethnicity! It's ridiculous." And it was ridiculous, within the set of definitions we have constructed in sociology that permits us to see ethnicity as culture, as community, but leaves out the body.

Ethnicity has a bodily component, and race has never been only about the body. The confusion of the social with the physical, the cultural with the biological, has been with us from the beginning. To invoke race, to discuss the different "types" of people on the planet has been to note the physical, the social, the cultural differences. The very concept of race was a cultural invention, a tool, and as John Edgar Wideman says, perhaps it has most profoundly always been a weapon: "The discovery of people unlike themselves did not spark in Europeans a doctrine of cultural relativity. It produced the invention of race. Of all the weapons devised to conquer and subjugate in the lands beyond Europe, the most effective, pervasive and enduring, the one that served to coordinate, harmonize and intensify the effects of all other weapons, is the concept of race."

The Science and Politics of Race

Can there be a science of race that is not scientific racism? Is it possible to maintain the concept of race without being racist? I don't think so. It is tempting, when we look backward at the racist uses of science, to say that the problem lies with the science: that their science was contaminated by their racism, that their science wasn't *really* scientific. We can say that of Linnaeus, and we can say it of Nazi race scientists, and we can say it of whatever American racist science is currently capturing the headlines. But if a record is so consistenly dismal, it raises the possibility that the problem is built not into the method or into the answer, but into the question. There may be no way to ask about the inherent biologic differences between groups of human beings without constructing race, and no way to construct race without invoking racism.

The apologists for racist science say we're afraid of the answers: we're afraid of finding out that there are real differences, particularly in intelligence, between groups of people. But no, it's not the potential answers that frighten me: it's the question. That question requires us to divide people into separate, ostensibly biologically based groups, and then compare them. Every time we start going down that road we get into trouble.

Is a better science going to produce a better set of facts and solve our problems? It has never worked that way before, and there is no reason for it to work that way now.

Racism cannot be argued away with facts. Part of the extraordinary power of racism, as with all ideologies, is its ability to incorporate any variety of facts and make them fit the belief. Darwinism offered the "survival of the fittest," an idea that both reflected the culture of the moment and swept that culture as an organizing principle. In the 1920s Darwinian thinking of this sort powered the eugenics movement. After a fashion. Consider how the facts were seen. Louis Terman, one of the early developers of IQ testing, in a typical statement of 1920s eugenics thinking, expressed his concern with where America was headed: "The fecundity of the family stocks from which our most gifted children come appears to be definitely on the wane. . . . It has been figured that if the present differential birthrate continues, 1,000 Harvard graduates will, at the end of 200 years, have but 56 descendants, while in the same period, 1,000 south Italians will have multiplied to 100,000." Which proves what? That south Italians (a "race" grouping of concern to Terman in that era of immigration) are more fit than are Harvard graduates? No, far from it. The "fitness" of the Harvard graduates is not open for discussion: these are the most gifted children and the source of all future most gifted children. They are "naturally" more fit than south Italians. So something is interfering with the natural order. Hence eugenics, a scientific practice to fit the scientific theory, to encourage the breeding of Harvard graduates and the limitation, through immigration laws, of south Italians.

Americans tend to think of eugenics as some strange Nazi pseudoscience, maybe something Hitler thought up. But eugenics was a solid, completely respectable part of American and British biology, taught in our universities, incorporated as the most up-to-date science in our laws. Immigration laws set quotas favoring those of "better stock," people from northern and western Europe, while restricting the "inferior" people of the rest of the world. Eugenics laws were also passed governing the behavior of American citizens: marriages of "feebleminded" people were restricted in the majority of states, but as an early German book on "racial hygiene" in the United States lamented, those measures were not implemented as rigorously as the eugenics laws in 32 states that prohibited marriage and sexual intercourse between blacks and whites. A widely distributed film of the silent-movie era, *The Black Stork*, advocated infanticide for "defective" newborns. It showed, among others, people with

Down syndrome as monsters best destroyed at birth. Children under the age of three with Down syndrome were the first group the Nazis gassed.

By the mid-1920s, an article about the German racial hygiene movement was able to state that there were virtually no differences between the position of eugenicists in the United States and in Germany, but that Germany lagged behind in terms of legislation, because "Germans are more disposed toward scientific investigation than toward practical statesmanship." A 1925 German pamphlet in favor of eugenics but cautious about compulsory sterilization claimed that in Germany, unlike in the United States, the right of self-determination was too strong to allow for the adoption of eugenic principles. Of course by the 1930s Germany had gone far toward eugenic policy implementation, and in 1934 the secretary of the American Eugenics Society observed, "Many far-sighted men and women in both England and America have long been working earnestly toward something very like what Hitler has now made compulsory."

It is less that the ideas were different under the Nazis; more that the targeted populations were different.

Back to Holland:

To get from our neighborhood out on the edge of the city to the center, we'd bike along a major thoroughfare. Roughly halfway there was a monument set alongside the road: four enormous hands, each larger than a person, rising, reaching upward. This was a memorial to the Jews of the city, deported and killed in the Holocaust. We pulled over one day, parked our bikes, and placed little stones on the memorial, the Jewish way of marking our presence at a grave. And it was time to explain the Holocaust to my five-year-old. She's my third kid; I've done this before. Explaining wars and lynchings and genocides and enslavements and rapes—all part of the parenting project. All part of the death of innocence, better done at one's own hands than by someone else, by the television. But not easy. So I did it, again, with the words that try to soften it a bit: long ago, bad people—the "not us, not now" story. Victoria asked all the hard questions you come to expect—even the children? the babies? and what happened to the bad people? I know it is complicated. In Holland we are living in an apartment building filled with old people. Besides my kids, I'm the youngest person I've seen there by 20 years. Where were all those old ladies when their neighbors were being marched off? But I don't even try to answer that. The kid is *five*. Back to the litany of long ago. But not that long ago, and standing here in Holland, not the comforting far away it was when I told the story to my other children. I'm tired, heavy of heart. We pick our bikes up and are walking back to the path when Victoria says, "I'm glad I wasn't born then. I wouldn't want them to get me."

The black eyes are serious in the dark brown face, contrasting the cheerful colors of the beaded braids framing it. I've explained segregation, Jim Crow, slavery and the backs of busses. Now what am I supposed to do with this one? "Yeah, me too," I say, and we get back on the bikes.

This one, this time, it wasn't about *you*, I'm thinking. This time race sliced a different way: it was Leah and Daniel they would have wanted, my white Jewish kids, not my black kid. Hitler was none too fond of Africans, true enough, but Jews were the focus of Nazi racism. Americans think of Jewishness as a religion; the Nazis thought of it as a race. Victoria's Jewishness was not the kind the Nazis were interested in.

In a war that produced a lot of memorable photography, one of the pictures that sticks in my mind is of a black guy in a Nazi uniform, lounging against a tree. Men of African descent served in Hitler's army. There is something nonsensical about that to a lot of Americans. The Nazi race system was not the American.

The American race system goes by color, from the whitest to the blackest, with Jews being not-quite-white. It's hard to picture the white American country club that would discriminate against Jews but not against blacks. Our race system is a continuum with a certain amount of confusion about the ranking of the off-white ethnics, but a great deal of clarity that all the white people are whiter than all the black people. Nazi antisemitism wasn't about Jews being less white, closer to African blackness. It was about Jews being Jews. It was the Jewish *race* they wanted exterminated from the planet.

The current debate in Holocaust studies focuses on the question of uniqueness: how different was what happened in Germany from other genocides, from other antisemitisms, from other human nightmares? But however similar or different, however much it did or did not draw upon older Jew-hatred, the concept of *race* is central in understanding what the Nazis were doing and what they did. The killing of Jews occurred in a context of killing to heal, a biomedical program, a eugenic program. There was, in Germany as in the United States, a concern with forces that weakened the Aryan/white/dominant race. While the United States contained its eugenic program to anti-"miscegenation," sterilization and immigration laws and a debate about infanticide, the Nazis went on to killing the "defective" and the "lives unworthy of life," a concept that expanded from individuals to races. The language was consistently a medical language, as Robert Jay Lifton and Erik Markusen point out in their history of the genocidal mentality, with talk of "racial pollution," "racial tuberculosis," Jews as a "cancer," "bacteria" or "parasites." "This medical imagery was no mere colorful metaphor: it was a social theory of collective decay, a social diagnosis containing a clear direction of treatment. As early as 1920, Hitler had made clear that the only solution for the Aryan malady was 'the removal of Jews from the midst of our people.' The vision then was not just biological but biomedical." There was a logical progression, if the word "logic" has any meaning here at all: from coercive sterilization, to the killing of "impaired" children, the killing of "impaired" adults, the selection and killing of "impaired" inmates of concentration camps, and finally the mass killing of Jews. All of these steps were "expressions of biological purification, of destroying 'bad' genes and 'bad' racial elements in order to revitalize the Nordic race and the world at large."

To have such a program, to make such a progression, there has to be a concept of race, a separate biological group capable of infiltrating, weakening, destroying not by military force, but by procreation, by the force of life itself. For the Ku Klux Klan, blacks were such a force; lynching was the cure. What needed to be protected was the purity of the white race, the purity of white womanhood. For the Nazis, it was the Jews, and Aryan or Nordic purity was at stake. In racist states—in America, South Africa, Germany under the Nazis—it is the power of race to live, to procreate, to produce and reproduce itself that must be contained. In such states, race is a contaminant, threatening to spread and take over. When a Jew and an Aryan come together and make a child, the child is a Jew under antisemitic racial logic; when a black and a white make a child, the child is black, in contemporary American racial logic.

American racism has centered on people of African descent, but other "inferior racial stocks" were also of concern in the passing of eugenic immigration laws in the United States. Laws were passed about Jews, but also about the Irish and Italian races. Nowadays speaking of Jews as a race sounds antisemitic. And speaking of the "Irish race" or the "Italian race" sounds just plain weird. When I argue that there is no such thing as race, that race is a socially constructed category of convenience and not a biological given, lots of people will nod sagely and agree with me. But if I were to refer to the Jewish race or the Irish race, people will indignantly, angrily, harshly inform me that these people do not constitute "racial" categories. Is there no such thing as race, and especially no such thing as a white ethnic race? If speaking of Jews as a race is itself racist, antisemitic, if asking a Jew to check off "Jew" in the race category of a questionnaire is inappropriate, offensive—well, what then of "White, Black, Asian and other"?

Americans are increasingly unsure of what it does all mean. Race is in legal terms a "suspect category," and increasingly it is a suspect category in conversation. Because race has so regularly been invoked when it is entirely inappropriate, we hardly know how to name it when it is appropriate.

I sat on a panel at a conference some years back. There were about six or seven of us up on the stage, some men, some women, a few blacks, mostly whites. I was wearing a bright red dress, as was the black woman sitting next to me. Someone in the audience wanted to ask one of us a question, to address it specifically to one of us on the panel, but couldn't remember our names. She sat way in the back, and gestured rather hopelessly at our end of the panel. Her question was "for the woman in the red dress." The black woman and I looked at each other, back out at the questioner. Another awkward moment of fumbling. "Which one of us do you want," I asked, "the white one or the black one?"

I could have let her off the hook more nicely of course—I could have said "Me or her?" What I did was rude, violated American race etiquette—we presumably aren't supposed to see it. The red dress was a legitimate thing to notice, as for that matter was our assumed gender. But the eye that could notice the color red was not allowed to acknowledge brown and beige, black and white.

People certainly do see race. We see race as this physical reality, this recognizable pattern of differences between people. It is foolish to try to persuade people that the differences don't exist. They do. It is pointless to try to convince people that the differences don't matter. They do.

What confuses us is that the differences exist physically, but matter socially. There are physical differences, and even physical consequences. But there is not a physical cause-and-effect relationship between them. Take something relatively simple: There is a much higher infant mortality rate among blacks than among whites in America. The differences between black and white women are there, real and measurable. But those differences, the physical, biological characteristics marked as race—level of melanin in the skin, shape of the nose or whatever—are not the cause of the different infant mortality rates. The darkness of the mother is a physical, biological phenomenon, as is the death of the baby. But the relationship between the two is a social reality: it is the social consequence of race that causes the physical reality of the death.

When we try to study race this is the trap we fall into: trying to construct a science of race that starts with groups of people and then looks for the differences between them, we end up trying to "explain" difference. And since the difference we began with is a difference we claim as "pure biology," we are hard put not to see its consequences as equally biological. That is the trap that the controversial 1994 book *The Bell Curve* fell into. First the population is divided into "black" and "white." Then the differences between the two, in this case the differences in "intelligence," or more specifically IQ test scores, are measured. You explain away all the difference you care to with other factors: whatever is left is race itself. Or so it seems.

Explaining wasn't the concern of the earliest sciences of race: cataloging was sufficient. The differences were strikingly apparent in the early days of "contact," as European explorers surveyed the globe and its peoples. Race was a way of naming the profound otherness they experienced—and, too, a way of profiting by that otherness.

That concept of absolute otherness is hard to maintain. We are not separate species. Race moved inward as blending occurred, from the skin, the body, the dress, the "obvious," to blood. Race was a liquid concept: it would blend. But race was also an absolute difference; it became like the red sock in the laundry, a source of stain, of contamination, a violation of the whiteness and purity of whatever it touched.

That was the fascinating concept that allowed slave owners to be fathers of their own slaves. Those black children of their "blood" were not comparable to the white children of their blood: stained forever with their mothers' blackness, they were forever other. The Nazis claimed the Jewish children—or grandchildren, or great-grandchildren—of Aryans to be Jewish. A "one-drop rule" prevailed in the American South, with any black "blood" making a person black. With more sophisticated accounting, the Nuremburg rules of the Nazis established a one-sixteenth rule.

Blood has ceased to be the powerful metaphor it once was. Essence has moved further inward, from blood to genes. No longer visible, no longer divisible, race has moved inward from body to blood to genes; from solid to liquid to a new crystallization. Blood no longer tells. Race is now a code to be read; the science of race is the science of decoding. The discriminating, discerning, trained eye that can recognize the "essential" or defining characteristics in the individual that confer race categorization will now be looking not at the face, the angle of the cheek, the color of the skin, but at the DNA.

Through a Crystal, Darkly

When you start thinking about people as compilations of finite numbers of bits, just so many allelic variations of just so many genes, then the idea of "related" is transformed. If two people have the same allele, that is, the same form of the same gene, then they do; if they don't, they don't. History is irrelevant. It doesn't matter where they got it from, and you can't tell no matter how carefully you look at it. The gene might have come from the germ cell of one to grow into the body of the other, as parent to child; it might have come to both from the germ cell of a shared parent; it might have mutated spontaneously in one or both. It is either there, or not there, the same or not the same. My redheaded sister-in-law, Jill, does not share those alleles that produce red hair with her father. She does, however, share them with my stepfather, "Red." He was a redhead. She is a redhead. They each carried two alleles of this most recessive of hair colors. The genetic connection between them could go back to the first redhead, or to the various spontaneous mutations that produce redheads. Jill and Red were "relatives" by way of two separate marriages, my mother's second marriage to Red, and Jill's marriage to my brother. They are not "related" as we say "by blood." But there that gene sits, as shared as it could be between father and daughter, as shared as it could be between any two strangers on earth.

My daughter by adoption is the only one of my children to have "my" eyes, my father's eyes. We're myopic, nearsighted. My father had died by the time I got my first glasses. My eyes filled when I took Victoria for her first pair, when I saw that pleased smile as she took in all the newly sharpened detail. I remember so clearly the way that the pile in my mother's grey rug differentiated itself again into separate tufts with each prescription change, blurring over the year to a flat grey expanse before the next appointment. Victoria looked down and smiled, and I saw her eyes through my eyes, through his eyes.

Is that a gene? Might be, might well be. Did I give it to her? Of course not. Do we share it? Yes, we do.

Victoria is missing a couple of tooth buds for adult teeth; the baby teeth are there with nothing coming along to replace them. (I look forward to major dental bills on that one.) I mentioned it to my great aunt Eleanor: "Oh, that's a Colb trait," she told me, naming the various cousins on my grandmother's side who were missing adult tooth buds here and there. Had Victoria been born to me, we'd have seen that as coming from my side, given to her by me. Again, wherever this trait comes from, it is shared by the people who have it. Bit by bit, across 23 sets of chromosomes, across perhaps 100,000 gene-bits, we match or do not match each other, person by person. If this missing-tooth trait turns out to be a single gene variation, then she does have this "Colb" gene, no less so for not being a Colb by birth, by "blood."

My children by birth share traits with me. They look like me. The odds of shared traits with my children by birth are certainly higher than the odds of shared traits with my child by adoption. As children of my eggs, we share a lot of traits, but on any given trait, they either do or do not share an allele with me, and express it as I do or do not. On any given trait, I am no more related to either of them than to the child by adoption. Bit by bit, point by point, gene by gene, we either share or do not share the allele. Is relatedness just a score card, how many points match?

Since 999 out of every 1,000 DNA base pairs are shared, we're all, all the people on earth, quite closely related. As of course we can see—any human being looks more like another human being than like anything else on the planet. So now I'm having a very hard time not getting all sloppy and sentimental about the human family, and how all children are the children of us all. Which, while true enough, doesn't seem actually to get us anywhere.

Victoria, after all, is black and I am white, and in America, toothbuds and myopia are nothing compared to melanin.

Some genes come to define people, to place them, while others are lost in a vast sea of individual variation. The genes that mark people as being of African descent are socially powerful in America, just as those that mark people of Maori descent are powerful in New Zealand. The particular angle of nose that marks a man as a Maori in New Zealand has no "meaning" in the United States. It moves into the realm of individual variation, of "family noses" perhaps, but not racial signifiers. A Maori woman found, "In Paris or New York or Amsterdam I was exotic. . . . It wasn't until I came back home again that I became just another Maori Sheila."

The traits that we think of as marking "race" are, as evidenced by the sight of "mixed-race" people, not single gene traits. Hair texture, nose shape, and skin color shades vary on a continuum. Each is the product of several different genes. It is that ability to mix, to blend, that gave rise to the idea of race as liquid, in the blood, flowing, staining and blending.

When race was liquid, was in the blood, it blended, stained and—maybe—washed out over time. For the Nazis, 15 parts Aryan were sufficient to wash out the Jew; for the Americans, one discernable, traceable drop of blackness irrevocably stained. A parent, a grandparent, a great-grandparent—some of their blood flowed in your veins, marking you.

When race was liquid, it could be placed outside, elsewhere, dammed off with "them." There is a pure, and a contaminant. You could start from anywhere, of course, and talk about contaminating. "Master was in your grandma's cabin just like mine," decries claim to African purity. In a white-dominated society, race is seen to lie in color, and whiteness is defined as the absence of color, as the purity of the unstained.

It doesn't actually work that way genetically, a thought which I find pleasing. It is whites, it turns out, that have the "race"-coded genes for skin-color variations, not the darker-skinned people. We all, black and white, have the melanin-producing genes. In lighter-skinned people, what *makes* lighter-skinned people, is a combination of genetic contributions that more or less turn off melanin production.

It is pleasing because whites, and particularly white men, have thought of themselves as "standard issue, nearest to God, the measuring rod for comparing other kinds of people. . . . Race was difference. . . . For all intents and purposes, whites viewed themselves as raceless," just as men viewed themselves as genderless, just ordinary, basic people. And yet it turns out that just as femaleness is the "natural" or "fallback" state of the embryo, that it is maleness that requires elaboration and differentiation, so it is with darkness and lightness. It is whiteness that requires differentiation, whiteness that carries race in the skin, whiteness that has the "race" genes. If we are made in God's image, then it seems she is, after all, black.

Wherever the characteristic or trait lies, crystallizing race, moving it from blood to gene, means that that which is white or black or Jewish or Maori or whatever is not distributed in blood and infinitely divisible. Rather, each trait is the product of one or more discrete lengths of code. There are a few characteristics, produced by a finite number of genes, that are differently distributed in different populations. That is what we see as race. Each individual either carries, or does not carry, the particular allele, form of the gene, that produces a given "race" characteristic. It is there, or not there. It doesn't fade out over time. It is passed on as a germ cell divides and unites to produce a zygote, or it is not passed on. Not liquid. Not blending. Crystal. Present or absent.

The human genome consists of something like 100,000 genes, each made up of many, sometimes many thousands, of base pairs. Each gene has various forms or alleles. Almost all of us share almost all of them: we are extraordinarily alike, all of us people of earth. Most of the variation is individual, within local populations: the differences within Swedes, within Maoris, within Greeks or Koreans, differences between individuals within a group. Less than 1 percent of the differences between people are the "racial" differences, whatever that could mean to us: the differences of skin and bone and hair, the blood group differences, the genetic disease variations, the frequencies that mark populations.

The differences that distinguish populations are finite. When counting the whole genome, there are only so many differences, and not many at that. At a population level, you can talk about percentages of shared traits, distribution variations across a continuum. Within the individual, there is no continuum: each individual either does or does not have any given allele. That is true of "racial" markers and it is true of "family" characteristics. While in one sense the body is deeply historical, there is another way in which it is bizarrely ahistorical. If you don't carry a particular gene, you don't. It doesn't matter if your parent did or didn't carry it. If he or she did not pass it on to you, you do not have it. Not a trace, a shadow, a remnant. It is there, or not there. Things "run in families," true enough. But if it doesn't get passed to you, it doesn't run in you or your descendents. In the blink of an eye, in the split of a cell, it is over. That bit of history ended right then and there.

What complicates the story a bit is that you cannot always tell which genes you carry: not every gene expresses itself. Some are recessive. You carry them and pass them on and never experience their consequences. Red hair, the most recessive of the hair colors, works that way. It slips silently along families of brunettes, only to burst into flame when two red-hair alleles join in a single child. And some genes are weak, not powerfully penetrant, unlikely to have an effect unless a number of other things, genetic and/or experiential, occur. These genes too slip along, only occasionally expressing themselves.

The liquid model saw characteristics as poured into children; in the new crystallization, "children are assembled as a collection of discrete, randomly assorted, stable, dominant and recessive ancestral alleles." Any given allele is there or not there, included in the assembly or left out of the package. Name the genes that most essentially define race—or any other aspect of the parent—and they can be gone in the next generation, simply not passed on, not included in the assembly. Another bit can be carried along quietly, not expressing itself for generation after generation. Yet another bit can change, can mutate, can spontaneously take on a new form. A "genetic" trait can be passed on in a family—or can arise spontaneously. When the gene is there, it is there, wherever it came from. When it is not there, it leaves no trace, no shadow, no touch of relatedness.

When race was solid, it stood. When race was liquid, it flowed. When it is crystal, it twinkles, on and off, bit by bit, fragmenting.

(American) Racism

Ideas, unlike genes, do carry historical shadows. We're not ever going to be able to think about race independently of its history. The concept always carries with it its own historical baggage. And just because there are no biologically specific determinants of race, no objective scientific markers to code race categories, that does not mean that there are no historically specific determinants of race, no cultural ways to code. A code is no less precise or objective because it is culturally rather than biologically specific. Consider the rules of spelling: They are real, identifiable, can be built right into your spell-check program. Thar ar rite an rong wayz too spel. And within a given culture, people can learn those rules and apply them objectively, systematically.

To read, you need to know not only the language, but the spelling rules. *Huis,* pronounced "house," means "house" in Dutch. Gaol, pronounced "jail," means "jail" in British English. It is altogether arbitrary, yet consistent within the culture. Race is a culturally and historically specific code, no less real and no less powerful for not being scientifically or biologically or genetically codable. To read race, you need to not only recognize the distinguishing characteristics (the letters, the features) as culturally defined, but learn their culturally specific meanings, recognizing not only the letters or features, but the rules that govern them.

As I've been writing this section on race, I've given it to a British woman who's been in the States for more than half her life, to a white American woman married to a Jamaican man, to a Japanese American woman who spent World War II interred in an American concentration camp, to a Haitian American, and to a Dutch woman who spent six months studying in the States. What I learned from each of their readings is that basically what I am saying is too American, it doesn't translate.

And how could it? Race *doesn't* translate—that is just the point. It is always constructed in historically and culturally specific ways. American racism is not German racism is not Dutch racism. It is not even the same racism attached to different races, although that apparently can happen too, as certain themes tend to repeat themselves.

BACK TO NEW ZEALAND:

I like local history museums. So in a quiet, midweek, midafternoon break in my schedule, the woman hosting us from the university took us to a wonderful museum of New Zealand history and culture. There were enormous woven mats, towering carved poles, the intricate and sometimes comical, sometimes scary faces and animals of Maori art. And there were artifacts, things and stuff, of New Zealand history, Maori and Pakeha (their word for the European settlers).

We wandered through the echoing space. And then our host, a very British-identified woman whose parents had come to New Zealand bringing English Christmases intact—goose, puddings, fruitcake and all—into the summery December of the Southern Hemisphere, began to explain her theory of New Zealand history and what was going to happen next. Maori sovereignty was the thing to watch. They were poised to take over. It was starting with the health-care system. She was a nurse, and she knew. Maori were taking over health care, and then they were going to take over everything. She spoke in hushed tones, partly out of the inevitable hush that falls in such a museum, but also it seemed the hush of paranoia. I was not (am not yet but most assuredly was not then) well versed in the politics and economics of New Zealand, but it was definitely my impression, I told her, that the Maori were poor. Yes, they are, she brushed that aside, but they are perched to take it over, they are poised to take over the whole country. They have a plan.

Something started buzzing in the back of my mind; this was familiar, but I couldn't quite get the name of it. I wandered away for a minute, trying to remember a title. Her voice followed me: The Maori, they're clannish, they stick together, they're like the Jews that way. Some people say they are one of the lost tribes of Israel. Ah, got it: "Protocols of the Elders of Zion." I'd approached a display on immigration to New Zealand, one of those mixed-media things: press a button and a voice comes out of the ceiling, the person whose photo and family memorabilia are in the case in front of you speaking from overhead.

I found myself before a display of Holocaust survivors, looking at the photo of a woman who came to New Zealand from Auschwitz. Just about far enough, I guess, given the size of the planet. I pressed the button, heard the voice of a Holocaust survivor describing hell, saw the earnest expression of our host explaining the future of New Zealand to Maren, my traveling companion, saw the Maori carvings through the doorway in the distance.

One form of British racism, like the British sparrows, seems to have neatly transplanted itself.

America, like New Zealand, is a country ruled by immigrants. To understand our racism(s), you have to know a lot of history. The essential, defining racism in America is of course the black/white divide, Africa on one side and Europe on the other, each defined in terms of the other. In American thinking, if European is culture, African is nature; if European is refined, African is primitive; if European is master of the world, African is its slave.

But that is not by any means the only racism we carry. There is a film I used to show to undergraduates, in which Bill Cosby sits on a chair, his face a painted mask to make him Everyman, and recites a litany of racist name-calling, all kinds of racial groups vilified. For a solid half-hour. Any middle-aged American can follow it. I had to stop showing the film—too many immigrant and young students in my classes had no idea what he was talking about part of the time, and it kind of defeats the purpose if you have to explain wop, kike and mick.

It is less the content of racism that concerns me here, as I think about genetics in the American construction of race, than its form. Wily, stupid, drunk, crafty, inscrutable, oversexed, clannish, brutish—the specifics vary. There are many different stories. What makes them all *racist* stories is that the characteristics of individuals are seen as bred into them as members of racial groups. In older imagery it was in the blood; now it is in the genes.

Racism isn't the only form of evil in the world. Hitler and the Nazi scientists understood the Jews as a race: the characteristics that mark Jewishness were inbred, and people were killed because of that. The Argentinian totalitarian state more recently also killed and tortured its defined enemies. But the Argentinians took the babies and the young children of the disappeared and raised them as their own. Hitler, in contrast, passed laws to ensure that no Jewish babies escaped: he had all abandoned newborns killed.

Some antisemitism historically permitted Jews to convert, to give up their Jewishness. That will work if the essence of the self, if the Jewishness of the self, is seen to lie in the soul. Racist antisemitism can have no conversion escape clause—you can't escape your body, your blood, your genes, can't convert by an act of will or an act of faith.

It is the belief in race as an immutable, inborn characteristic of the individual, its body-, blood- and gene-based essence, that brought the Japanese Americans and not the German Americans into concentration or "internment" camps during the Second World War. American citizens, including those born in the United States who had never been outside of this country, were rounded up and taken out of their homes, off their land, permitted to take only what they could carry, and placed in concentration camps. Because they were "racially" Japanese. This wasn't about "loyalty"—young Japanese men fought in the American army while that same army detained their relatives in the camps. That is the nature of the "race" concept—that being Japanese, or Jewish, or black or white is in one's essence, irreducible, an enduring truth of identity. Being German, on the other hand, was understood in terms of loyalties, chosen identities, nationality and citizenship. German Americans might have been pressed to prove loyalty, but Jewish Germans were inherently and inevitably Jewish, and Japanese Americans were inherently and inevitably Japanese.

These then were some of the racisms that America took into World War II: In a war against racism, eugenics and a "master race," Native Americans were enlisted off of reservations, black Americans rode segregated trains to segregated training, and Japanese Americans went from internment camps to boot camps.

Of course this is hard to translate.

For Whom the Bell Curves

I want to talk about the African American experience of racism and I find myself trying to do a calculus of anguish. I want to say something about the—what?—unique? persistent? widespread? special *somehow* variety of racism aimed at people of African descent in the United States. And every word I try to use to distinguish that particular racism starts an argument in my head. Remember what happened to the Japanese. Think about Chinese American history. Antisemitism slammed escape hatches from the Nazis. Remember the smallpox-infested blankets distributed to Native Americans. And think about Mexican Americans.

And yet, I need some way to distinguish this very special form of American racism from all the other forms. Because this, the black/white issue, *is* race in America; this is the color line, this is *the* issue.

Discover, a science magazine aimed at the educated, interested lay public, did a special issue on the science of race in 1994. A black adolescent boy naked to the waist, in a rural setting, holding a soccer ball, illustrated the table of contents; a picture of a black barber shop, black kid in the chair, black barber behind him, illustrated the editorial. The first white person whose illustration I found was a line drawing of Linnaeus. The content of the articles couldn't have been more antiracist, more discrediting of racist science and arguments. But the illustrations made it clear that race is *about* blacks, *about* people of African appearance.

For white Americans, race enters a room when a person of color comes through the door, just as for men, gender enters with women, just as for able-bodied people, (dis)ability rolls into the room with the first person in a wheelchair. The racelessness of white is its most distinguishing feature in America. Coming here, all kinds of people shed their race: the Jews became white, the Irish became white. Asians didn't but they are perhaps, as someone once put it in regard to international adoptions, a discrete shade of off-white.

To talk about race in America is to talk about people of African descent. And sometimes it feels as if talking about *anything* in America is talking about race.

That is certainly the lesson of *The Bell Curve*, both as a book and as an incident, an event, a chapter in the American story. By now the dust has settled on that, and America is probably in the quiet period between major, best-selling racist tracts. Every now and again one pops up—a book or an article that says out loud what people have been unwilling to say but apparently perfectly willing to think: that there are deep, permanent, inherent differences between "the races," and that a more scientifically objective social policy would recognize and incorporate that insight.

The Bell Curve, in case you missed it, said that in this great and open society of ours, intelligence is what counts. For people who have not got intelligence, "life gets worse, and its members collect at the bottom of society. Poverty is severe, drugs and crime are rampant, and the traditional family all but disappears. Economic growth passes them by. Technology is not a partner in their lives but an electronic opiate. They live together in urban centers or scattered rural backwaters, but their presence hovers over the other parts of town and countryside as well, creating fear and resentment in the rest of society." Intelligence is what counts, and intelligence lies in the genes: "Putting it all together, success and failure in the American economy, and all that goes with it, are increasingly a matter of the genes that people inherit." And the kicker, if the description of people at the bottom was too subtle: Intelligence is measured by tests which show that "the average white person tests higher than about 84% of the population of blacks and that the average black person tests higher than about 16% of the population of whites."

The book spawned a small industry, a veritable genre of critiques. The methodological critiques, taken together with the book, would make a terrific course in social science methodology: how to do it and how not to do it. I am not going to give you a summary of those many arguments here, but the book is just irresistible—no one can resist taking a few shots at it. But more important for my purposes right now, to understand the way that genetics as an ideology functions in its role in American racism, nothing could be better than *The Bell Curve.*

The book explains that blacks are poor because they're, well, black. Lower IQ scores are a part and parcel of blackness, a piece of being of African descent, and from that comes poverty, social disorganization, bad mothering, gang formation, and whatever other social ills you can name. That is a genetic explanation of race and racial stratification. That explanation effectively removes racism from the equation: blacks aren't poor because of anything structural at the level of the society, but because of structures in their own brains, their own wiring, coming from their own genes. This argument says we need not look to racism to explain what is happening to blacks in America. The explanation lies in race itself.

The authors of *The Bell Curve*—and I really am taking them as straw men, handy representatives of racist thinking alive and kicking in racists all over this country—think that the idea that there is a serious racism problem in America is, to use their own word, "implausible." They have made the argument that IQ is at an individual level 60 percent inheritable (a point of considerable methodological criticism) and that there is a standard deviation of 15 IQ points that typically separates blacks and whites in test scores in America. To explain black/white differences in IQ by environment would mean, following just the mathematical logic of this, that the mean environment of whites is 1.58 standard deviations better than the mean environment of blacks, "when environments are measured along the continuum of their capacity to nurture intelligence." I don't know how you could or would measure an environment's ability to nurture intelligence. But if you think that this means there is a very racist environment, one in which black people are systematically mistreated physically, socially, psychologically, that seems reasonable to me. Lower IQ in blacks is what we in sociology call "overdetermined." From nutritional deprivation through lead poisoning through the social consequences of poverty through lousy medical and educational services, there are more than enough explanations—not even discussing the ability of the tests to measure intelligence. To *The Bell Curve* authors, "environmental differences of this magnitude and pattern are implausible."

That point is a key argument in the ideology of genetic determinism: environmental differences can't possibly matter very much. You can acknowledge that there *are* environmental differences, but their effects are trivialized, minor variations in a symphony played by genes.

Counterarguments lie right in the data that the authors present in answers to questions they pose as rhetorical. They note that the environmental effects of racism are not distributed equally along class lines, and ask, "Why, if the black white difference is entirely environmental, should the advantage of the 'white' environment compared to the 'black' be greater among the better-off and better-educated blacks and whites?" They announce, "We have not been able to think of a plausible reason." I can explain it in a five-minute visit to my local public school, filled with poor kids, mostly black with a smattering of poor whites as well. I find it a very depressing place, and along with everybody else with any money in my neighborhood, have bought my kids out of it.

If instead of assuming that some people have "smart genes," one assumes instead that most people are capable of becoming very smart indeed if placed in an environment that "nurtures intelligence," then the question they ask is pretty answerable: poor whites are not that much better off than are poor blacks. Poverty hurts all children, diminishes their intellectual capacities. Being poor is not good for one's intellectual development in America, land of private medical care and local funding for schools. Poverty has physical, psychological and social consequences that place a ceiling on intellectual development for many of the children who experience it. Is there a clearer way of saying that? Poverty sucks. It sucks the smarts right out of kids, black and white.

Better-off whites buy better environments. Better-off blacks cannot entirely buy their way out of the effects of racism. Ellis Cose in *The Rage of the Privileged Class* talks about the racism that exists and is experienced even at "the top" for American blacks. White Harvard professors treat black students differently than they do whites— whatever makes one think that the elite kindergartens that feed Harvard are any better? So of course the differences will show up more at the economic top than at the economic bottom.

I do wonder just what *The Bell Curve* and other genetic determinist race arguments think the environment is. At best they take a very narrow, simplistic view of "environment." Consider how they dealt with the widely cited Minnesota study of transracial adoption. Now granted, I'm more than a little touchy on this subject. I was reading *The Bell Curve* late one night when Victoria came down the hall from her room to crawl into bed next to me. I looked from her sleeping face to this disgusting book and put my head down and cried. But I don't think this is about *my* reading of the data: I think it is about *theirs*.

The Minnesota study showed, at age 7 (around Victoria's age now) only slight support for "genetics," only slight differences between the white and black adoptees. Those slight differences are to be expected, given the environmental consequences of racism during their birth mothers' lives to the point of pregnancy (like lead exposure, nutritional deprivation, health problems); the environmental consequences of racism experienced during the pregnancy itself and in whatever time elapsed between birth and placement (usually longer for black children), and the troubling but possible existence of some racism in the adopting households themselves. But at a follow-up study ten years later, with the kids in adolescence, race made a big difference. As the authors of *The Bell Curve* see it: "The bottom line is that the gap between the adopted children with two black (biological) parents and the adopted children with two white (biological) parents was seventeen points, in line with the Black/White differences customarily observed." In an older, pregenetic language, this would have been phrased as "Blood will tell." But what I learn from this data—and what it terrifies me to be reminded of—is that white parents can give their black children some of the privileges of whiteness, but only in the earlier years, when parents generally have more control over their children's environment. As our children grow, they grow out of our homes, out of our carefully selected, edited, manufactured world, and into America.

And it really is America that produces this. The authors of *The Bell Curve* use the Minnesota adoption study to support their genetic causality argument, rather than as I do, to refute it. But they do acknowledge that not all studies on race lead to the same conclusion. A box on the same page as the adoption study reports,

> One of the intriguing studies arguing against a large genetic component in IQ differences came about thanks to the Allied occupation of Germany following World War II, when about 4,000 illegitimate children of mixed racial origin were born to German women. A German researcher tracked down 264 children of black servicemen and constructed a comparison group of 83 illegitimate offspring of white occupation troops. The results showed no overall difference in average IQ.

German racism is not American racism, and black IQ isn't their issue.

The variety of racisms in the world is not apparent to the authors of *The Bell Curve.* That book naively assumes that any racist system is interchangeable with any other, and ought somehow to have the same consequences. "An appeal to the effects of racism to explain ethnic difference also requires explaining why environments poisoned by discrimination and racism for some other groups— against the Chinese or the Jews in some regions of America for example—have left them with higher scores than the national average." Antisemitism, the specific form of racism that has targeted Jews, was never about Jews as beasts of burden suitable for slaving in fields,

Jews as primitive, undeveloped, animalistic creatures of the jungle. We're the crafty ones, remember, the ones that are going to take over. And the Chinese—the inscrutible Oriental? Central casting sends us the wise old Asian and the foolish black: Detective Charlie Chan is not to be confused with his eye-rolling, foot-shuffling, "feet don't desert me now" black sidekick.

It is naive—it is silly—to say that if one kind of racism doesn't appear to lower IQ then other kinds of racism won't.

Before I finish with *The Bell Curve,* one more issue remains. (And when I finish, what do I do with it? Literally, the book itself? It feels like a particularly ugly pornography, whose First Amendment right to publication I defend with all my heart, and whose physical presence I need to protect my children from. Maybe I can give it to a graduate student studying research methods.)

The issue that remains is race itself: if you're looking at differences between "the races" in intelligence or anything else, you have to have a category of race and a way of sorting people by it. The authors don't:

> We frequently use the word *ethnic* rather than *race* because race is such a difficult concept to employ in the American context. What does it mean to be "black" in America, in racial terms, when the word black (or African American) can be used for people whose ancestry is more European than African? How are we to classify a person whose parents hail from Panama but whose ancestry is predominantly African? Is he a Latino? A black? The rule we follow here is to classify people according to the way they classify themselves.

But that self-classification is limited to the categories of white, black, Asian and occasionally Latino. In their definition, "white" becomes one ethnic group, comprising Irish, Jewish, Italian, Greek and whatever other "white" Americans. Black is another ethnic group, including southern Americans or South Africans; and Asians a third, including Chinese, Japanese or Indian and Pakistani Americans. Whatever.

Earlier I raised ethnicity as a way of understanding the differences in physical appearance between peoples without invoking race. Ethnicity is a useful concept when read *against* race, not when used as a synonym for it. But in a racist society, nothing escapes a racist spin. What might have been a relatively harmless, even helpful concept of ethnicity also gets caught up in a racist analysis. *The Bell Curve* just uses the word, as if it were perhaps a more polite way of talking about race, less "difficult" in the "American context." (Race is such an ugly word.) Other uses of ethnicity have been just as troubling. If there are no races but only ethnic groups, then what explains the enormous variation in "success" between various ethnic groups? Why, that nonbiological argument goes, did the Jews, Irish and Swedes all "make it" while the African Americans continue to languish in poverty and despair? For those who don't want to place the answer in biology, an alternative is to place the answer with culture.

And so a myth of cultures of success and cultures of poverty as causes rather than consequences of difference is developed. Some cultures or ethnic groups are poised for success, some for failure. Blame it on child rearing, on mothers, on attitudes toward education, on whatever you choose. See what distinguishes the successful from the unsuccessful and claim those differences as the causes of the success or failure. This is the process that William Ryan identified as "blaming the victim," a confusion of cause and effect, results identified as causes, outcomes used as justifications. This is a real danger of using ethnicity as a way of understanding difference.

Just because there aren't "races" doesn't mean that there isn't racism. That is the distinction my colleague and former student Heather Dalmage is making when she says, "If we change our language to *ethnicity* we are not disrupting traditional understandings of *race* as a biological condition." In a racist society, it is absolutely necessary to retain "race" as a political concept. Descendents of African slaves in the United States might very well form an ethnic group, as might descendents of Swedish laborers. But that doesn't make the experiences of the two groups comparable: one faces racism and one does not.

Race remains both a meaningless category, and one of the most meaning-laden categories we have for people. The biologists and geneticists may be more than ready to let go of race as a category of analysis—but the rest of us cannot afford to, not as long as racism exists.

Rates and Races

I've found my sociological training, my sociological imagination, surprisingly useful and appropriate in studying genetics. That probably sounds contradictory—most of what we read and hear about genetics counterposes it to sociology as different ways of understanding. Things are either nature (genetics) or nurture (sociology). Why do little boys play with trucks? Something on the Y chromosome? Or something on the TV?

So certainly there are oppositions between the two ways of thinking. But there is something profound that resonates between sociology and genetics. Both make sense at the level of the population in ways that they will never make sense at the level of the individual.

The first sociological study was Emile Durkheim's analysis of suicide. Just over a century ago, Durkheim compared suicide rates between different groups to understand the phenomenon of suicide and to understand something about the way the groups were organized. From where we are now, all of us at least somewhat sociologically literate, that seems reasonable. But just think about it: he was trying to explain something that was so intimate, so personal, so deeply individual, by comparing social groups. It was a magnificent intellectual breakthrough.

If Finns have higher suicide rates than Italians, what can that tell you? It cannot tell you anything about why any given Finn kills himself. No one, I should think, kills himself *because* he is Finnish. And it can't tell you which Finn is going to kill himself and which not. But it does tell you that the annual rate of suicide among Finns and among Italians is surprisingly, remarkably stable, year in and year out. The *rate* is a characteristic of the population. No matter how high the suicide rate is, it is still the case that hardly anyone kills himself. The chances of any given person killing himself go from really remote to really very remote as you compare one society to another. At the level of the individual, knowing the suicide rate for the population is fairly useless: it doesn't allow prediction and it doesn't aid explanation. At the level of the whole, at the level of the population, though, it does both of those. Studying suicide rates can enable you to predict the rate of any given country for the coming year, and to predict whether that rate will go up or down within that country given certain social changes.

If you are interested in the varieties of human experience, you can learn most of what is interesting within any given society. You can find Italian suicides and Finns who overcame every conceivable adversity without thoughts of suicide. The interesting, extraordinary and complex differences between individuals exist within the population. But if you want to study the population, if you want to study Finnish or Italian society, you won't get there by reading suicide notes. You'd do better to look at the rates of suicide—and of births, marriages, crimes, drug use, alcohol use, literacy, etc.

So it is with genetics. There are differences between populations and there are differences among individuals. As I discussed earlier, most of the range of human variation can be found between individuals within any given population. And what does distinguish populations is mostly useless at the individual level for prediction or for explanation.

Variation in blood groups, for example, exists in pretty much all populations. If you want to do a transfusion, you have to type for A, B and O, and will find people of each type all over the planet. And yet population differences do exist. Given the earlier association of race with blood, an enormous amount of research went into distinguishing populations by blood-group frequencies, with early claims to distinguish "European" blood. It played out in a more complex way. Differences can be found in ABO frequencies between populations, but one is hard pressed to find meaning there. As Jonathan

Marks points out, "A large sample of Germans, for example, turns out to have virtually the same allele percentages (A=29, B=11, O=60) as a large sample of New Guineans (A=29, B=10, O=60). A study of Estonians in eastern Europe (A=26, B=17, O=57) finds them nearly identical to Japanese in eastern Asia (A=28, B=17, O=55)."

The ABO frequencies are a characteristic of the population. If there is a disaster in Estonia and you are in charge of bringing in blood supplies for transfusion, knowing the distribution of ABO among Estonians is valuable information. If you are a doctor faced with a bleeding Estonian, the information on distribution rates is useless. And if you are a researcher with a vial of blood, knowing its type won't tell you if it is Estonian or not. Rates, whether of suicide or of blood groups, are characteristics of populations, not individuals.

That has two consequences, at two levels. One is that you cannot learn about individuals from groups, or about groups from individuals. They are separate levels of analysis. As a sociologist, I've learned to think about prediction at the level of the whole, and sometimes at the level of segments of the whole—but I've also learned *not* to try to make predictions at the level of the individual. I think in terms of rates. So I can think about the relationship between, say, social class and health, and know that the higher people are in the socioeconomic system, the greater the chance of their being healthy and of having healthy children. And at the same time, it doesn't surprise me one bit when a wealthy person dies young, when a rich and powerful person loses a child to disease, or when a poor, uneducated person lives to a hundred having raised ten healthy children and scores of healthy grandchildren and great-grandchildren. I don't think of any of that as contradicting what I know about the relationship between class and health, because what I know is about relationships, correlations, rates, and not individual causality.

That is why it makes sense to me that Estonians and Japanese have relatively high rates of type B blood, and it also makes sense to me that the vast majority of Estonians and Japanese do not have type

B blood. The rate is high relative to other rates; the individual instance is low.

One of the more satisfying ways we think about social class in sociology is in Max Weber's terms: as "life chances." Anything can happen to anybody, but your chances of certain kinds of things happening vary with your class position. The poorest kid in the most miserable neighborhood with uneducated, unemployed parents can grow up to win the Nobel Prize in physics, and the richest kid from the wealthiest suburb can end up sleeping in doorways. But that doesn't take away from the fact that the first kid's chances of success are different from the second kid's, just as a Japanese kid has (slightly) greater chances of type B blood than does a German kid. That sociological perspective is useful for reading genetics. When I read, for example, that Ashkenazi Jews have a higher rate of a gene associated with breast cancer, I read that as a sociologist: I note the association, and I note the implications for life chances, and I also note the limits of that for individual prediction.

The second set of consequences of understanding rates as characteristics of populations is that you can, usefully, study populations *as* populations. They can't be reduced to the level of the individual, but populations *can* be studied at their own level. What can we learn when we study population genetics, rates of genetic distributions within populations?

When we talk about groups of people and their genes, what immediately comes to the (American) mind is *race*. As race moved inward, from the body to the blood to the gene, genetic differences became the measure of race. But race, biologists and anthropologists tell us, has no meaning; it doesn't exist scientifically, objectively.

Race has shaped the lives of people on earth in every possible way. An enormous number of people have suffered, died, because of race. How could race be so perfectly obvious to see, so enormously powerful in its consequences, and yet not exist? What could it possibly mean to say that something *that* real isn't real?

Race implies discontinuities in the differences among people around the world, and race also implies some essential differences.

To sort people by race is to use one category and assume others, to mark race by one trait and then expect to see other traits sorted along with it: to start with skin color and find IQ; to start with hair type and go to rhythm. No single trait means race: they have to cluster into categories by which people can be distinguished, or what have you?

Jonathan Marks, whose work in this area I've become so fond of, gives a good example to explain some of this. If you had a big pile of blocks of different sizes and asked a child to sort them into "large" and "small," the child could do it. If a number of children attempted that task, there would be complete agreement about some, maybe most of the blocks—the largest and the smallest. "The fact that the blocks can be sorted into the categories given, however, does not imply that there are two kinds of blocks in the universe, large and small—and that the child has uncovered a transcendent pattern in the sizing of the blocks. It simply means that if categories are given, they can be imposed upon the blocks."

Sure we can sort people by, say, skin color, and get considerable agreement, at least at the extremes, about which people go into which group. But skin color, like block size, is a single trait. What if some of the blocks were plastic, some wood and some metal? The child could sort by that and get a completely different arrangement of blocks. And if some of the blocks had grooved edges and some smooth? The child could sort by that too, with yet another arrangement. And if the blocks were painted red, yellow and blue? You could sort blocks any way you want to, but if the color, size, grooves and material are not systematically related to each other (if not all grooved blocks are larger than smooth blocks, or all yellow blocks metal) then you're going to end up with entirely different piles depending on what you sorted for.

People can be sorted by color. But they can also be sorted by any number of other characteristics. The presence of a skin fold on the eye. The presence of loops, whorls or arches in their fingerprints. The presence of specific genes, like the sickle-cell gene. The ability to digest lactose. Each sorting will give you different "race" groups, just like each sorting of the blocks will give you different piles. And no amount of sorting will tell you about the essential types of blocks, or people, in the world.

When you sort people by skin color you get some people from India sorted with some people from Africa—the way my Dutch neighbors saw it. Sort by eye fold, and one group of Africans goes in the same race as the Asians, and the rest of Africans go with the Europeans. Sort by fingerprint type and most Europeans and most Africans go in one group, with Khoisans and some central Europeans in another and Mongolians and Australian aborigines in a third race. Sort by sickle-cell trait and Nelson Mandela's people, the Xhosas, get classified with the Swedes while most African groups get classified with the Greeks. Sort by lactose tolerance and northern and central Europeans go with Arabians and the west African Fulani, while most of the rest of the African groups go with the east Asians, American Indians and southern Europeans. Sort by ABO blood groups and the Germans go with the New Guineans and the Estonians with the Japanese.

Do you think you've actually discovered some important characteristic of the Estonians now that you know that? Is there some indication that perhaps, had that fact been known, Estonians would have been in camps with Japanese during World War II and the Germans would have declared the New Guineans of Aryan blood?

Skin color and various cranial measurements were abysmal failures in classifying people into demarcated racial groupings. Blood typing raised hopes that science could get to the bottom of it, could accurately sort people by an objective criterion. But that didn't work either, with each new factor in blood typing serving only to further muddy what distinctions could be made.

With molecular genetics, and its ability to see differences not in the body or in the blood, but inside the cell, new hopes were raised— here we could locate race, locate the fundamental groupings of people. And here too confusion reigns as the genes that mark "race" twinkle on and off all over the world.

There aren't any markers that set people apart into race groups.

There are any number of characteristics that can be used to sort people, that can be imposed, like "large" and "small" on the blocks, but
none that emerges as an inherent difference. There is too much genetic variation within populations, and too little between them, for
"race" to make sense. Race can and will continue to be imposed on
human groupings, but the clusterings that exist do not display any
discontinuity that makes a boundary or boundaries discernable.
Those boundaries will always be in the eye of the beholder.

Does that mean that we can't learn anything about humans
from the ways that genes do cluster in populations? Of course not:
the first thing we learned is that there aren't any races. The various
groupings of people both separated out too recently in human evolution and have had too much continued interbreeding to sort out
into distinct groups. We vary all right, but not in massive, structured, systematic ways.

The other thing we may hope to learn is something about our
history. Individual genes carry no history: they are present in one allele or another, with no narrative about how they got there. You can't
read history from a single genome. You don't really need genomes to
understand that. Think about a "mixed-race" child, say one who is
"half" of African descent and "half" of Asian descent. That child has
two grandparents of African descent and two grandparents of Asian
descent. Looking at the child, you cannot tell which grandparent
mated with which: did an African couple produce one of the child's
parents and an Asian couple the other? Or did two Afro-Asian cou-
ples produce two Afro-Asian parents who produced that child?

Can clusters of genomes, patterns of genetic distributions, tell us
something more about history, more about how they got that way? If
two groups share some characteristic, some gene, with each other
and with no one else, we can ask how that came to be. Like looking
at the Afro-Asian child, we can read a number of possible stories
from that. An attempt to read those stories, to gather a narrative of
human travel through the planet from our genetic clusters, has been
attempted. That is the goal, or one of the goals, of the Human
Genome Diversity Project.

The Human Genome Diversity Project

It was the Human Genome Diversity Project that inspired me—or if inspiration refers to giving breath, perhaps inflamed would be more accurate—to write this book. I spent a weekend in November of 1995 at a conference on the human genome at Stanford University. One day was devoted to the genetics issues I have long been concerned with, issues usually subsumed as "bioethics." The other day was focused on the Diversity Project.

I embroidered: the world going to hell in a handbasket. I got the handbasket with the globe in it done, and spent the plane ride home figuring out how to represent hell in colored thread.

It was, as has often been my experience of genetics conferences, an Alice-in-Wonderland kind of thing. We were carefully and repeatedly reassured that genetics research will show us clearly, once and for all, that there is no biological basis for race classifications. And a man who was identified as the only Native American Ph.D. geneticist in the world was introduced, I kid you not, as a "full-blooded Hopi."

The Human Genome Diversity Project presents itself as an attempt to read human evolutionary history from our DNA. The plan is to get DNA samples from 25 individuals from each of several hundred groups of indigenous peoples around the earth, and from that try to tell how we became who and what and where we are.

At the time of that meeting at Stanford the project was already well on the defensive. It had been attacked by virtually every concerned party in every imaginable medium. Geneticists and anthropologists, indigenous people on the Internet—all had had a go at it. And, as is often the case, its supporters claimed that all these attackers just didn't understand. To know us, they said, would be to love us. The project will prove, by looking at difference, how similar we are.

Neither I nor most of the social scientists in attendance were able to see how any listing of shared genetic history, when combined with unique genetic markers, was going to bring us the end of racism. Conduct this little thought experiment: If Hitler had won, if current human genetic research were being conducted under the auspices of the 50th anniversary of the Third Reich's world dominion, and swastikas rather than stars and stripes hung in the halls of Stanford, would that kind of research be safe? Do you think that perhaps if we could have shown Hitler how very slight the Jewish genetic indicators were, he would have said, "Never mind"?

It is most assuredly not that the impetus for the Diversity Project is all bad. It addresses some very real limitations of the Human Genome Project. But as I spent that day listening, seething and embroidering, I kept thinking, and occasionally asking, are you people crazy? Do you have no idea of what you are doing and risking? This is safe only to the extent that the world we now live in is not the world that Hitler would have created. It is safe only to the extent that racism is gone. Not too safe from where I sit.

But first let me take a step back. What is the problem that the diversity project will resolve? One problem is claimed to exist in the world itself: indigenous people are dying out or breeding out. Groups that the project founders called "isolates of historic interest" are heading toward extinction or dispersion, and information will be lost. If they don't die off altogether, their genes will be mingled and lost to study. Cultures, language, skills, knowledge and *people* are being lost: genetic information is not high on everybody's priority list as indigenous people the world over suffer the consequences of worldwide capitalist exploitation.

The other problem to be solved is the problem with the Human Genome Project itself. Let me first address that, and then turn to the criticisms of the Diversity Project. The Human Genome Project is often said to be "mapping the human genome." But since, first, it is not a map, and second, there is no such thing as "the human genome," that is misleading. To the first point, the project is constructing not so much a map as a transliteration. Recall that DNA is constructed of paired molecular discs called base pairs. There are four kinds of molecules organized in two kinds of base pairs: one constructed of adenine and thymine, and one constructed of cytosine and guanine. If you abbreviate these as A, T, C and G, and "read" the base pairs as they exist on a stretch of DNA, you are transliterating or transcribing from molecular forms to letters. Given how unimaginably tiny these molecules are, it is quite a technical accomplishment. But it is "reading" only in the sense of naming the letters. A child is said to be able to read her name when she has learned to associate the name of each letter with its shape on the page (see adenine as the letter A) and when she can recite the letters in the order that makes up her name (see the order of the base pairs as they exist along a stretch of DNA). At that point she's not even sounding it out: it is simple transliteration. This squiggle has this name, the next squiggle has that name, and this is the order of the squiggles. She has no idea what it means or what would change if the letters were rearranged.

So it's not a map, but a sequence or a transliteration of the human genome. But that makes it sound like there *is* a human genome, like there's a name BARBARA that can be read, or a downtown Brooklyn that the third grade can go out and map. And that's the second point. To the extent that we're all the same, such a map can be drawn. To the extent that we differ, it can't. We are the same in 999 out of 1,000 base pairs. That is a lot of similarity. We are different in 1 in 1,000 out of the total 3 billion base pairs. And that is a lot of variation.

What is being transliterated is most often described as a composite or a consensus genome, as if one could construct a "generic" person. Think about the drawings you have seen of the human skeleton: those too are a form of "map," a line on the piece of paper representing the outline of a bone in a body. Whose body? Just any old body, a generic body.

Bones are easily counted, and their points of connection easily established. The hip bone is connected to the leg bone and all that. Perhaps some day the genes will be as well mapped as the bones are, even if there are a vastly larger number. But that is not where we are now. Less than 10 percent of what is being transcribed is estimated to be genes, the stretches of DNA that code for specific proteins. They don't even know yet just how many of those genes there really are. One hundred thousand is an oft-quoted estimate and it has been the one I've used in this book, a nice round number to work from. The other 90 percent of the DNA exists between the genes, and has been dismissively referred to as "junk DNA." The majesty and mystery of DNA has been reserved for the part that is being decoded as genes.

Where a gene is understood, what will be mapped is the gene in its "normal" form as the reference point. For example, the allele that doesn't cause dwarfism will be considered standard, rather than the one that does. Choices will have to be made all of the time, just as when you construct a drawing of the human skeleton. Some of us have a middle toe that is longer than the big one: which type shows up on a "standard" skeleton? And who decided? Certain political and ethical issues make themselves immediately apparent: which do you think will be the standard, the allele that is believed to "cause" homosexuality, or that which is believed to "cause" heterosexuality?

Most of the choices are made without a clue as to what they are choosing between. Since they don't know how many human genes there actually are, it's hard to say what percentage have been identified. But with a lot of rounding off, we can call it less than 10 percent of the genes which are themselves perhaps 10 percent of the DNA. That means that less than 1 percent of the whole long string of ATCGs can be "read" in any meaningful way at all. It's almost as if someone who couldn't read but did know the alphabet picked representative parts out of a bunch of names from a phone book to get a "composite" name. What is being sequenced or transcribed is just a representative or composite or consensus genome. But what or who on earth does it represent?

The people most enamored of the Human Genome Project like to refer to the genome as a "blueprint" for a person. That makes it sound like one could follow the plans and construct the person whose genome will be mapped. Who would he be? And he would indeed be a *he*. For reasons which are sensible enough based on the way that stretches of DNA have organized themselves into chromosomes, the sequences will include a sample or representative of each chromosome, and that includes one X and one Y chromosome, as well as the ones numbered 1 through 22. So our representative generic person, with a Y chromosome, will, one more time in our history, be a male. That alone should make it clear that there is no generic, representative, standard model of the human, and that every choice that is made has political implications.

The other way to go, if you don't want to focus on a single, representative human, is to try to get at the diversity of human genomes. But what exactly would that mean in this context? Most of the variation between people is to be found within any given population. But most of the variation that we think of as diversity is the variation between populations. When I ask for more diversity in a classroom, I am not asking for a wider assortment of random alleles. I am looking for "racial" or "ethnic" diversity. When I look for diversity in a neighborhood, I might also be looking for variation in age, class and forms of household living arrangements, all things (I believe) unrelated to genes and all forms of diversity (I think) are important.

Diversity, then, is always in the eye of the beholder. I recently heard a radio report on the establishment of Levittown, an early post—World War II planned suburb. One of the first people living there remembered it as a wonderfully diverse community: and then began to name all of the various jobs held by the young white heterosexual married men with children who bought these houses. From his perspective, if an advertising executive lived between a plumbing contractor and a social worker, that represented a wonderfully mixed community.

Because of the way that we think about genes, and because of the way that we think about race and ethnicity, when we ask for genetic diversity, what we are looking for are genomes selected from people of identifiable racial/ethnic groups.

If all of this were a purely descriptive project, sampling from genomes that appear all around the world would both make sense and be a benign activity. If you don't do that, you're in the make-believe land of one-size-fits-all and flesh-toned Band-Aids. But a map is never just a description, and there is no way of describing human variation around the world without invoking the history and the politics of race. That is why it is not safe, and this is perfectly demonstrated in the history of the Human Genome Diversity Project.

Begun in 1991 by two geneticists, Luca Cavalli-Sforza and the late Alan Wilson, the Human Genome Diversity Project was to serve as a corrective for the limitations of the Human Genome Project, collecting samples of genomes from the many and varied peoples of the earth. It would also, according to its planners, tell us "who we are as a species and how we came to be." It is in that sense no less grandiose and no less limited than the Human Genome Project itself. The past, like the future, apparently lies in our genes.

The Diversity Project can be attacked, and is being attacked, on its merits, on its own terms, for its technical and theoretical shortcomings. And I will, shortly, discuss some of that criticism. But my focus here is on genetic thinking; the frame that genetics is offering us for viewing the world. And so once again the question isn't whether they're right or not, whether you can indeed learn the history of our species this way, but why you would want to. It reminds me of the light bulb joke: "How many New Yorkers does it take to change a light bulb?" The answer, I can tell you as a good New Yorker, is "Who wants to know?" Who wants to know, and why, are the important questions to ask here. And who wants to know is not only the researchers themselves—it is also who will use this research, who will draw upon it, and for what purposes.

If there were no power at stake, if there were no profound power differences in the world between groups of people, then systems of classification, by genomes or by anything else, could be used for their pure historical and inherent interest. Show a little child finger-prints, for instance, and watch how they explore the variations be-tween people in fingerprints. It's interesting, and why not? But there *is* power, and race is not a system of classification, it is a system of op-pression. And in such a system, there are grave dangers to introduc-ing classificatory categories of people.

One major area of concern addressed by the indigenous people themselves, the targeted populations, had to do with the finances. The 25 to 35 million dollars that the project was anticipated to cost pales of course in comparison to the 3 billion dollars for the larger Human Genome Project. But 30ish million dollars is a hefty sum of money by any other standard, and the politics of human diversity being what it is, "isolates of historic interest" tend not to be the wealthiest people in the world. Even within wealthy nations, it was poor minority groups that were targeted. For example, one group suggested for study were the "Etas" of Japan. That, now a pejorative term in Japan and no longer used, designated an urban outcast group, but one that was hardly genetically isolated. Which pretty much goes to prove that isolation is at least as much an economic as a genetic category. To spend millions of dollars sampling the genes of poor people so that they can be preserved for future study, while the people themselves are not being preserved, raises obvious con-cerns.

But the financial issues go way beyond the costs of the project. Taking genetic samples for research on biodiversity, and doing it in the more remote regions of the world, is not new. What was new was taking the samples of *human* biodiversity. The Rural Advancement Foundation International (RAFI) has long been concerned that the United States is patenting genetic plant material taken from many parts of the world. It was RAFI that first alerted the World Council of Indigenous Peoples to the Human Genome Diversity Project— starting a communication on Native-L, a First People's news net-work that alone makes the entire world of Internet communication seem worthwhile. To go to those same people from whom plants had been removed and patented, and take human genetic material, raised all the same concerns: would they now patent that? And it was not an outrageous concern. As Donna Haraway tells the tale,

> Inescapably, independently of the HGDP [Human Genome Diversity Project] but fatally glued onto it, the all-too-predictable scandal happened. . . . The Guaymi people (of Panama) carry a unique virus and its antibodies that might be important in leukemia research. Blood taken in 1990 from a 26-year-old Guaymi woman with leukemia, with her "informed oral consent," in the language of the U.S. Center for Disease Control in Atlanta, was used to produce an "immortalized" cell line deposited at the American Type Culture Collection. The U.S. Secretary of Commerce proceeded to file a patent claim on the cell line.

RAFI stepped in, informed the Guaymi General Congress, and the situation was brought up before the Biological Diversity Convention in Geneva. The U.S. Secretary of Commerce withdrew the application.

When a fortune of money is being spent to take cell samples from the poorest people in the world, asking "Who benefits?" is not unreasonable.

But finances aren't the only concern. The threat looms of biological warfare, a targeted attack by governments on portions of their populations. If the point is to find something "distinctive" about particular indigenous groups that sets them apart from later settlers in that same area, then can that distinction be used against those people? Remember the smallpox blankets that were given to Native Americans in the earliest known example of biological warfare. I don't think anyone thinks that that is the intention of the planners of the Human Genome Diversity Project. But how reassuring is that? In human history, information has been used in ways that its gatherers had not intended, and it can happen again.

Context is everything. A 1997 research project compared the Y chromosomes of Jewish men who claimed membership in the priestly class with other Jewish men, and found "distinctive genetic traits," indicating that "these men have the Y chromosome of Aaron." Judaism is a religion, but it started as a tribe. Membership, or Jewishness, as Jews traditionally define it, is passed from mother to child: any child of a Jewish mother is Jewish and any child of a non-Jewish mother is not born into the Jewish community. But within the community, or within the tribe, descendence is traced on the male line: people are named as the son or daughter of their father. Aaron, the older brother of Moses, was the priest of the tribe. His sons are called *ha'kohen*, which means "the priests"—giving rise to the contemporary Jewish names like Cohen, Kahane, etc., though some of us without those names claim to be *ha'kohen* descendants. My mother tells me my father was one, which would have meaning for my brother but not for my son.

The Y chromosome is unique among chromosomes because it is passed on intact: there is no maternal chromosome with which to recombine. So each man has his great-great-great-great-great-great ... grandfather's Y chromosome. Neat, huh? And in itself, a kind of harmless thing to know.

But let's give it a moment's thought. Within the Jewish community, if people started taking this seriously, genetic testing for *kohen* status (which has certain rights and obligations associated with it for the very religious and observant) would uncover issues of paternity. Since Jewish law permitted a man to divorce a "barren" wife, one can imagine that there was a certain amount of informal management of male infertility. Or, perhaps umpty-ump generations ago some man lied, for whatever reason. Say he was a *kohen*, but wanted to marry a divorced woman, forbidden for a *kohen* but permitted for others. Generations later these various lies would surface. All very awkward, at best.

And what about in the larger context, outside of the community of Jews? Who cares, you might well ask. But what if this technology was 60 years older than it is? Hitler's Nuremberg laws set the definition of a Jew as someone who is one-eighth Jewish, anyone who had one Jewish great-grandparent. What if Hitler could have had cheek scrapings done, getting cell samples, and identified every man—and his sisters—who were direct descendants of this tribe within the tribe of Jews? Or race of Jews, as Hitler saw it. Now my brother is vulnerable, and through him me, and through me, my children, even my son who does not have a Y chromosome from me. This marker goes back countless generations: they found it in the Ashkenazic Jews of European descent, and the Sephardic Jews, a split that goes back to a diaspora more than a thousand years old. Think how many people could be involved in more than a thousand years of passing a chromosome along. An intriguing little finding of historical interest in 1997 would have meant untold deaths in 1937, as millions more people would have been identifiable as Jews, carrying this marker, or as daughters or sisters of those marked.

Another marker, a mutation as apparently meaningless as a birthmark on your belly, has been found on the Y chromosome of many Native American men. Will it someday be the case that a man who wants to claim his rights as a Native American will have to submit to genetic testing? Genotype is already perceived of as more "real" than phenotype, an indication of who or what you *really* are, rather than what you seem to be. In turn, phenotype, or the being before you, is seen as more real than the culture or community to which that being claims membership. So membership in communities risks being reduced to genetic markers.

I pray that they don't find a marker that distinguishes the Serbs from the Bosnians. I thank the powers that be that the Y chromosome is small, and can't contain as many markers as there are vulnerable groups.

The Human Genome Diversity Project has echoes of the problems with *The Bell Curve:* not only is it very bad politics, but it is also very bad science. And it is the same kind of bad science, if somewhat more sophisticated and subtle. It is a science based on a presumption of difference that then shows difference.

Bad science is not necessarily in and of itself a problem. Sometimes it just occupies scientists, taking them down some wrong paths, and may even have the advantage of leading them to an interesting place. We do learn from our mistakes, oftentimes more than from our successes. But this is bad science with bad politics, and that is always troubling. Testing all the men of earth for what is presumed to be Aaron's Y chromosome would tell us something interesting about how people have moved around. Testing Jews for it tells us about Jews: and who wants to know, and why?

The methodological criticisms of the Diversity Project can be summed up, if perhaps oversimplified, by saying that it treats human population diversity as if it were species diversity. Which is much the same criticism leveled at classic racism and racist science too. Humans are all one species. Birds are not.

The interesting thing about species is that they do have their history written all over them. It's not easy to decode, but it is there. It is there when you observe the phenotype, as Darwin did, and it is there when you observe the genotype, as contemporary geneticists do. When characteristics or genetic markers are shared between species, that tells you something. It might mean that they separately evolved the same characteristic, but it also might mean that they had shared ancestry. This is the insight that Darwin had. What makes species readable historically is that they are indeed species: the hallmark of a species is that it cannot interbreed with another species. There are amazingly diverse breeds of dogs, but they can interbreed. There are some dogs that look more like pigs or cats than they do like most other dogs. But their dogness is determined by the fact that they can breed with other dogs, not with cats or pigs.

So when you find some set of characteristics shared between several species, you can know that that characteristic, unless it evolved separately in each, evolved in all of them back at the time when they were not separate species. Speciation is a way of categorizing, but it is also a moment in time—even if a moment counted in thousands and thousands of years—the point when two populations diverged so greatly that they become separate species, incapable of interbreeding. On the other hand, when you find some set of characteristics shared within a species by some but not all members, you won't know, cannot know, when it got there. If you find this Y chromosome of Aaron's in some men in some little island in the Pacific, it could be because one of the men who originally settled that island was a descendant of Aaron, or it could be that such a man or group of men arrived two hundred years ago. There is no point at which the flow of genes had to stop.

The Human Genome Diversity Project is founded on the idea that there *are* genetic "isolates," groups of people who are mixing genes only with each other and not with the rest of the world. The project operationalizes the concept of isolation by setting the date at around 1500, and saying aboriginal populations are those that were "in place" before then. It's an interesting model of the world. First we split off and wandered the planet, then we settled in place for a long, long time of isolation, and then Columbus arrived and things got mixed up. But as Jonathan Marks points out, "prior to Columbus there was Prince Henry the Navigator; prior to Prince Henry there was Marco Polo; prior to Marco Polo there were the Crusades; prior to the Crusades there were the Mongols and Huns; prior to the Mongols and Huns there were the Romans; prior to the Romans there was Alexander the Great. And those are just the European highlights."

We use the model of the Polynesians setting off in canoes and landing on isolated islands, like the Maori in New Zealand—as if the movement of people was a one-time dispersal outward to isolated posts. But if you take a good look at a globe, we don't actually have the separate continents they taught us to name in elementary school. You can pretty much walk the whole planet, and there's no reason not to criss-cross, retrace steps, wander around. The one thing we do know about people is that when populations of them interact, crossing each other's paths, little mixed babies are a sure product. So static isolation is probably not a good model for what happened to us as a species.

Maybe the problem is that the founders of the project were geneticists, rather than anthropologists, and missed everything that anthropologists have been saying in the second half of this century. In an earlier and more naive time, anthropologists went off and studied "tribes." In what was called the classic ethnographic period, from 1910 to 1960, the anthropologists typically went in for a year or so, checked out the kinship system as described, got the origin tales of the community, and observed their customs and practices. They worked with one or just a few informants and learned history. The history they learned was, it now seems obvious, just one history of that community, just one story. The actual messiness of life got overlooked: the ways that differences within the community played out; the ways that people from outside of the community married or bred in and people from inside married or bred outside; and the ways that origin tales had, as they always have, a political as well as a theological tale to tell. Most origin tales say, we were always here; we were the first people here; we are the true people of this land. More sophisticated contemporary field work has shown enormous amounts of diversity and "gene flow" between tribal societies once thought to be isolated.

If there aren't isolated aboriginal peoples, then sampling from them makes no sense: we're back to the disproved logic of race, in which you first divide the population into race groups, and then compare across race to discover racial difference. The Human Genome Diversity Project supporters vehemently decry the concept of race: they are interested in small populations, demes or clines or ethnos as they are variously called. But call it anything you want: if they are not intrinsically different and if they are not isolated, then you cannot learn history by looking at them now.

All of which is not to deny the obvious, the thing we come back to over and over again: people do tend to look like the people they live among. At the Stanford conference, Jonathan Marks argued that if what they are really interested in is the diversity in the population as it has spread itself around the planet, then put a big grid across the globe, and take samples from each box. It is a much safer way of getting that information, a way that doesn't risk endlessly recreating race categories even as we try to argue against them.

Communities, peoples, tribes, societies, groups of any sort, construct themselves *socially*. Reading membership from DNA reads against the text, substitutes "objective" or scientific information for the ways communities construct themselves. Consider this: genetically, I'd be just as related to some Cossack that might have raped my great-great-grandmother as to the woman herself. African Americans in the United States can trace their ancestry as much to any given owner who brutalized a young woman slave as to that young woman slave. As communities of people we have carefully crafted stories, histories of who we are, of what we choose to pass on and what we choose to let wash over us. These stories are not our ignorance speaking: they are our collective wisdom.

I have a small woven straw basket on my desk. It's a beautiful miniature of a rice-winnowing basket made by a Gullah woman living near Charleston, South Carolina. Africans who knew rice-growing techniques were captured from the area around Sierra Leone and brought as slaves to the geographically similar region near Charleston. Bits of language, fragments of songs, rice-cultivation techniques and these basket-making skills have survived in this community of slave descendants. A contemporary basket-weaver in Sierra Leone can recognize as one of her people's baskets a basket woven by an African American woman many generations removed from Africa.

You can read history in a basket, and you can read history in a cell. The relatedness of the two basket weavers might show up in their DNA as well, bits passed down generation to generation. Language and skills carry history; DNA carries history.

Could DNA be just one more way to read history? If it were, it could be relatively harmless, and kind of interesting. But we know, from our history, that when we face alternative tales—one told in our grandmothers' songs and stories, and one told in the language of science, that science takes on the authoritative voice, and becomes the *real* story, and the other becomes an "old wives' tale."

Genetic thinking places "authenticity," *real* membership in a community, in the cells of the individuals, not in the fabric of the people. Genetic thinking takes power away from people to tell their own history, and gives power to people with the technology to scientifically read that history. Science and technology, we like to think, are themselves morally and politically neutral, and we only have to worry about them if they are in the "wrong hands." Maybe so—

certainly science and technology in the wrong hands are very much worth worrying about, and nowhere is that more true than in the sphere of race. But are there right hands? Or doesn't the very existence of "isolates of historic interest" being studied to the tune of 25 to 35 million dollars suggest that questions of "race," of "identity," of "descent" can never be understood without the power relationships that created them?

Genetics didn't create racism. Genetics, not just as a science or set of technologies or practices, but genetics as a way of thinking about the world, maintains and supports racist thinking just as much as it has the potential to undercut and counter racist thinking. There is this endless interplay between an ideology, a way of thinking, and the science that is being developed. Scientific discoveries come out of ways of thinking, and scientific advances capture the imagination, give us ways of thinking. There isn't some pure science going on out there somewhere, producing facts which then get used to produce the way we understand the world.

Ours is a world in which some people are resources for other people. The same power relationships that "discovered" the dark continent and the new world, that brought back spices and gold and people, now brings back samples of plant and human genetic diversity. And—still—puts them up for sale. In a racist society, in a world in which race is a system of power and oppression, genetic thinking is going to be used to support that oppression. How much our science of genetics will support or will undercut racism won't depend on the "facts" genetics has to offer: it will depend on the world in which those facts are offered.

Part II

WRITING THE BODY: THE GENETICS OF ILLNESS

A Tale of Two Diseases

Popular wisdom has it that eugenics died out as a way of thinking and as public policy after the Second World War. Some say it is only now, with the new genetic technology, coming back in a new form, coming in "the back door," in Troy Duster's evocative phrase. I think eugenics never disappeared from American thinking. The word certainly got discredited, but the idea didn't. The idea is that nature cleans up its own mistakes, but on the one hand, the process is perhaps slower and less efficient than we may like, and on the other hand, modern living and particularly modern medicine are circumventing nature's intention: "We keep the genetic weaklings of our world alive and allow them to reproduce at will, to pass along to the next generation the same traits that nature in its inherent biological wisdom would have eliminated by killing off their owners long before they could have reached the age of reproduction."

That was basically the argument that the eugenicists of the 1920s and '30s were making. That quote, however, comes from a book based on a CBS television series produced in 1967, long after World War II, and before the widespread use of genetic testing. Walter Cronkite was the narrator and principal reporter. This was a mainstream American product. Fred Warshofsky, the science writer who wrote the book as well as the program, went on to say, "We have a softer view, a shorter view of life than nature does, and our so-called compassion in this regard becomes positively myopic when we regard what the biologists call the gene pool."

What exactly is it that we would do if we were truly compassionate and abandoned our "so-called compassion"? What is it we would do if our vision were clearer, less myopic? Would we kill off the genetic weaklings? Or is that perhaps where the Nazi program reached its excesses? Maybe we would permit the weaklings to live, but apparently we would not permit them to reproduce, not permit genetic weaklings to taint the gene pool. Forced sterilization was of course a key element in all of the eugenics programs, including in the United States, Germany and Sweden, a powerful tool for protecting the gene pool from contamination and deterioration.

And just what is this "gene pool"? In spite of all the vicious uses of the concept, I find something delightful in the *idea* of a gene pool, of all of the genes just floating around. I like to think of the human gene pool, all of its—no, all of *our*—glory and diversity, our tallness and shortness, lightness and darkness, hair kinky and straight, eyes blue and green and black, myopic and farsighted, just all of it floating there, endless potentiality, a vast resource on which we continually draw, making wonderful new combinations of people.

Others take a narrower view. The American and the German eugenics programs weren't about *the* gene pool because they weren't thinking about *the* human race. My idea of the gene pool is a public pool; theirs is a gated community. That's one way of thinking about the genetics of race: Race implies separate gene pools, and racism requires keeping them separate, protecting the gene pool of one group from contamination by another.

The language of liquid is so seductive. I'm not sure how much of it is left over from the language of blood, but it is such an evocative language. The geneticists speak of gene flow, the movement of genes between pools. Racists think of this as a contamination, but we can instead think of it as an infusion of fresh blood, a sprinkling of new resources. For all the the crystalization of genes, for all of their particulatization, it still works as a metaphor to think of them as suspended in liquid, pooled, flowing between people.

Macroeugenics is about keeping the pools separate, free of contamination by other, less valued pools. Macroeugenics is eugenics as Hitler perfected it, a eugenics of race.

Microeugenics is a genetics of disease, a eugenics we are still working on. Microeugenics filters each pool, whether considered as separate or as one grand human pool. Floating in there, along with all the delightful things, are the terrifying things: the genes that cause evil, destruction, horror on the body. Of course there is a politics here in picking and choosing between genes, which are "good" and which are "bad." But it's not all politics: a gene that causes physical suffering and early death is no one's idea of good. There are such genes and we rightly fear them.

That is the basic model of genetic disease most of us have learned: Floating out there in the gene pool, carried in streams by families, are genes that cause disease. Tay-Sachs is a classic model of this kind of disease, sickle-cell anemia another. Tay-Sachs disease has become a touchstone in genetics, both because of the horror of the disease, and because of the success of the screening program. Sickle-cell anemia presents another model, a model of everything that can go wrong in a screening program.

Tay-Sachs is a dreadful disease: a baby normal in appearance and development for the first few months, or even as long as a year—smiling socially, engaging with its parents, learning to sit, pull itself to standing, understanding language—deteriorates over a period of years to an inescapable, lingering death. The baby has seizures, loses its ability to respond, to understand, becomes unable to hear, see, move, finally unable to breath. The only treatment medicine had—or has now—is to prolong the dying. My mother used to work at what was then called the Jewish Chronic Disease Hospital, which had a Tay-Sachs ward. She was there the night of the infamous East coast blackout. She was a dental hygienist, and spent that night hand-suctioning the babies' throats. She got a commendation for her work, keeping Tay-Sachs babies alive through that night.

My mother is no believer in prolonging life unnecessarily. When I was working on a book about prenatal testing, I called her up and asked why she had done it, why she had participated in dragging it out longer for those babies. She didn't really know, was bothered by the question, had never really thought it through. The babies were just babies, thick with mucous they couldn't cough up, looking in that dark and quiet night like any other snuffly baby. So she moved back and forth all night, doing by hand what the machines usually did, keeping the babies alive to suffer through another day. "I couldn't just stand there and watch a baby choke to death in front of me." And who could, and who could blame her?

There were 20 babies, she thinks, on that ward, maybe more. That was 30 years ago—there are no such wards any more. Medicine couldn't prevent those deaths, or even prevent itself from preventing those deaths when they were due. But medicine has learned to prevent births. Screening, prenatal testing and selective abortion have prevented the births of those babies, a solution, troubling in its own way, that we will come back to later.

Tay-Sachs is what is called an "inborn error of metabolism." The baby with Tay-Sachs accumulates gangliosides, a type of lipid, in the brain because an enzyme that would permit their metabolism fails to function. That enzyme can be provided, but medicine hasn't found a way of getting it through the blood-brain barrier, to the brain where it is needed. The nonfunctioning enzyme was discovered first; the genetic error was localized to the 15th chromosome more recently. As with many things in life, the story seems to get more complicated the more it is understood. At first it was always described as a simple disease involving recessive genes: each parent could be a carrier, and if two carriers mated, the chances were 25 percent that the resulting baby would inherit both good forms of the gene, two normal alleles, and be fine; 50 percent that the baby would inherit one bad and one good allele and so be a carrier like its parents; and 25 percent that the resulting baby would inherit both Tay-Sachs alleles and have the disease. Little diagrams of this were regularly used in presentations on genetics to explain how it all works.

More recently I've started to hear at genetic conferences about parents who were on the "high end" of normal, not quite carriers: being a carrier became a relative rather than an absolute status. And then I started hearing about late-onset Tay-Sachs, a form of Tay-Sachs that shows up in adolescence. That simple on/off-switch model of a gene worked better when the gene itself was seen as a single unit. Once the gene is understood to be composed of thousands of base pairs, the possibility opens up for variations. Not all mutations are necessarily the same.

But Tay-Sachs remains the example of choice for people who want to push for more genetic screening and testing. It is the model screening program.

Tay-Sachs is not free-floating at equal rates throughout the gene pool. As most Americans have heard, it is a disease associated with Ashkenazi Jews, the Jews of eastern Europe. The chances of someone of Ashkenazi Jewish descent carrying the Tay-Sachs allele is roughly 1 in 30; the chances of a North American non-Jew carrying the Tay-Sachs allele is roughly one-tenth of that, 1 in 300 or maybe 400. Not only Ashkenazi Jews carry Tay-Sachs: the British inhabitants of western Newfoundland have relatively high rates, as do Cajuns of Louisiana. It is this bobbing to the surface of disease-linked genes that makes one think about the kinds of issues the Human Genome Diversity Project addresses. Did the disease start with a mutation in an Ashkenazi Jew? Did some Jewish wanderers bring it to these other places?

But with those questions come the troubling questions of race, of identifying Jews as carriers of a "Jewish" disease. Is there a way that we can think of Tay-Sachs as a "Jewish disease"? And what does that mean? The disease is no more common among more rather than less religious or observant Jews—it is no test of one's "Jewishness." Nor does it seem to correlate with other "Jewish" traits: some carriers "look" Jewish, some don't. But if you think about populations as groups that tend to keep dipping in the same gene pool, that breed more within the group than outside of it, then Tay-Sachs is a disease found more often in the Ashkenazi Jewish gene pool than in other pools.

There is an irony in using a "Jewish disease" as a hallmark of microeugenic programs. Jews were the highly targeted victims of the best-known program of macroeugenics. And there has been enormous sensitivity to that history and the concerns it raised. Tay-Sachs programs were established with consultation with Jewish leaders, and Jewish doctors were involved at every stage of planning. The histories fold in on themselves in such interesting ways: because of their own commitment as a religious group to marrying within the community, but also because of antisemitism, Jews constructed a relatively distinctive gene pool, walled from without as much as gated from within. Because of the commitment to reading and studying as part of the religious observation for Jewish men, but also because of antisemitic laws forbidding Jews to own land, Jews were well placed to take part in the explosion of education as a pathway for upward mobility. Occupations of the bourgeousie, including medicine, opened up to those for whom the peasantry was closed.

So what with one thing and another, we got the Jewish Chronic Disease Hospital hosting a Tay-Sachs ward, and Jews openly accepting one of the first major programs targeted at identifying genetic carriers of a disease.

The story of sickle-cell anemia is *very* different, as one might well expect. Like Tay-Sachs, sickle-cell anemia comes from a disease-causing mutation that floats at higher rates in some pools than in others. Like Tay-Sachs, carrier status appears to have some advantage: Tay-Sachs carriers appear to have some heightened immunity to tuberculosis, a serious disease problem in crowded ghetto conditions; and sickle-cell carriers have some immunity to malaria, a serious disease problem in mosquito-infested parts of Africa and southern Europe.

Sickle-cell is thought of as a black or African disease, although it is not equally distributed through African populations. Within the United States, people of African descent, as well as people of European-Mediterranean descent, Italians and Greeks, are more likely to be sickle-cell carriers. For African Americans, the rates are reported as between 1 in 10 and 1 in 13, but given the difficulty of defining African Americans under the "one-drop rule," all such numbers are debatable. A person whose ancestors are as likely to have come from Europe as Africa is declared African American, but is less likely to carry a sickle-cell allele than one whose ancestors are almost all from those parts of western Africa with the highest rates of sickle-cell disease. It is not ever possible to talk about or think about population groups without thinking historically and without thinking about racism. You might think you're talking about a biomedical phenomenon, but you're also talking about the rape of slave women by white masters looking to produce more property.

The sensitivity that was shown in setting up the Tay-Sachs programs was lacking in the sickle-cell programs. There is an ongoing debate about how much antisemitism remains in America, and so there is a potentially interesting discussion about the role of antisemitism on the one hand, and responses to anticipated antisemitism on the other hand, in setting up Tay-Sachs screening programs. The role of racism was a lot more blunt and far more divisive in the sickle-cell programs.

The early 1970s, when both screening programs were developed, represents very different historical moments for blacks and for Jews in America. Jews were well assimilated by then, generations removed from the bulk of immigration. Jews were in positions of authority in medicine, and Jewish communities were well placed to guide the development of the programs. Tay-Sachs screening was of, for and by the Jewish community.

Sickle-cell screening, on the other hand, did not emerge from the black community, though some groups did take it up as an issue. Civil rights activists were concerned and vocal about the dramatic black/white differences in health, but those differences were rightly understood to reflect the consequences of racism and poverty. The Johnson administration had declared war on poverty; Nixon declared war on cancer. When it was judged politically advantageous to do something "for" the black community, the administration targeted sickle-cell disease. Congress passed the Sickle-Cell Anemia Control Act in 1972. The very word "control" is enough to raise hackles.

States then began screening programs, but in a remarkably poorly planned way. Counseling services weren't available, and educational services weren't available, neither for the individual being screened nor for the public at large. This wasn't something a community was doing to help itself; this was a congressional program to control a disease in an already burdened community. There was mass confusion about the difference between sickle-cell disease and what was called sickle-cell trait. Sickle-cell, like Tay-Sachs, is a "recessive" disease: one needs to inherit two sickle-cell alleles, one from each parent, in order to have the disease. Carrying the allele for sickle-cell, like carrying the allele for Tay-Sachs, does not give you the disease. But sickle-cell carriers, unlike Tay-Sachs carriers, were said to have the "trait." And with very poorly documented evidence, it was said that some sickle-cell carriers under conditions of extreme oxygen deprivation showed some signs of sickle-cell disease. Sickle-cell disease and sickle-cell trait quickly collapsed in on one another in the public mind and airlines began grounding and firing sickle-cell carriers, the U.S. Air Force Academy excluded sickle-cell carriers, and insurance companies began denying insurance or raising rates. It was a total disaster.

And it was a disaster aimed at people who had no reason to be forgiving about such disasters. The Tuskegee syphilis experiments had only recently come to light: medical personnel, doctors and nurses, had deliberately left black men with untreated syphilis so that they could chart the course of the disease. There was also, in the 1970s, an increased awareness of the long-standing practice of forced sterilization for young black women. The civil rights movement and the women's health movement publicized what had been going on quietly, from the eugenics era of the 1930s through to the late 1960s and early 1970s. Cases became widely known of even very young black girls being sterilized without their knowledge, or their parents' knowledge. This was not a good context in which to come bearing the dubious gifts of carrier screening and later prenatal diagnosis and selective abortion.

The problems were ongoing. When I was doing research on genetic counseling and services in New York City in the early 1980s, I attended a meeting for genetic counselors who were on their way up to the state capital in Albany to lobby for funding. The goal was to get free genetic services available to the poor women of New York

City. An admirable goal: if there are genetic services available to people with money, I believe they should be equally available for those without. But if you're lobbying for funding, altruism and social justice are not your strongest arguments: money is. And for Albany, talk of "poor women downstate" means "black women." The man educating the genetic counselors on how to lobby told them to remind the state legislators that every sickle-cell fetus aborted would save New York State $20,000 a year.

If I wanted to make up two hypothetical cases of genetic disease to show the ways that race, culture and politics would effect the development of screening programs, I couldn't have done better than Tay-Sachs and sickle-cell.

It is not only the context and the people that were different. Sickle-cell disease is also a very different disease than is Tay-Sachs: people with sickle-cell disease have been known to lead rich, full, long lives. Unlike Tay-Sachs, the disease is more common, and the outcome so variable that people diagnosed as carriers might well know someone, have someone in their families, who had the condition and was doing okay, whose life did not seem to be a tragedy worth avoiding. While Tay-Sachs was known as always a very simple clear-cut case of absolute early death, sickle-cell disease was known from the beginning to be far more variable in its consequences. This variation in disease outcome itself tells us something interesting about genetic disease, about prediction and control.

Sickle-cell is a "point mutation." A single base-pair error results in sickle-cell: just one "misprint" in the three billion ATCGs, but strategically located near the beginning of the beta-globin gene. The protein that is produced as a result of this error is different from the nonsickling hemoglobin protein; it changes the shape of the red blood cells. They "sickle," or bend and so clog capillaries, potentially doing damage to blood vessels throughout the body. Sickle-cell disease is marked by "crises," times of acute pain, especially in the arms and legs, as the cells clog the capillaries. These crises can be rare or frequent; they can last for hours or weeks. The disease can kill early on, or permit a near-normal life span.

There are two interesting lessons to be learned here, about the limits of genetic understanding for prediction and for control. Ruth Hubbard explains:

> Scientists now know the precise molecular structure of the allele associated with sickle-cell anemia, and for several decades they have known the specific molecular change in sickle-cell hemoglobin that is responsible for this condition. Yet this knowledge has not enabled them to understand why some people who have sickle-cell anemia are seriously ill from earliest childhood, while others show only mild symptoms later in life, nor has it helped produce cures or even effective treatments. The best medical therapies for people with sickle-cell anemia still rely on antibiotics that control the frequent infections that accompany the condition.

Genetically, this illness is understood about as well as anything has ever been understood. That understanding buys you little in the way of prediction and nothing in the way of treatment. Why is one child in agony and another mildly affected? And what can we do to help the one in pain? Take two aspirin and an antibiotic and call in the morning.

Sickle-cell anemia programs have improved dramatically since the disaster of the 1970s. The treatment may not rely on genetic knowledge, but the treatments are more effective. And the worst excesses of the screening frenzy are over.

But between them, these two diseases, Tay-Sachs and sickle-cell anemia, set the frame for understanding genetic testing for disease. People, it turns out, are more complicated than pea plants, and Mendelian genetics doesn't look like it did in the book. And studying disease genetics or studying population genetics, we take our history with us.

The Quest

Some genetic diseases, like Tay-Sachs and sickle-cell anemia, are attached in our minds to "populations," to "ethnic" or "racial" groups. Cystic fibrosis, a genetic disease affecting the respiratory system, has been called "white folks' sickle-cell" or "goyishe Tay-Sachs"—it is carried by people whose ethnicity Americans generally erase, white Anglo-Saxon types, the "everybody else" when the "ethnics" are separated out.

But there's another way of thinking about genetic disease, not at the level of the community, but within families. Some families carry certain diseases, passing them on, generation to generation. Sometimes these are family secrets, hushed, never spoken of in front of the children. And sometimes these diseases are the focus of the family identity, the very thing that unites them as a family. Some of those families have been discovered by geneticists; some have called geneticists in. I've been reading about dozens and dozens of such families, stories both sad and uplifting, stories of fate met.

There is a developing genre within genetics narratives now that is a variation on the family saga: the story of the family curse. These curses are real, they are terrifying, and the courage and strength of the families that face them have all of my admiration. But when you've read the sixth or seventh in a row, you begin to get the plot line.

Probably the best known of these is the Huntington's story. I first heard of Huntington's chorea during the folk-music revival of the 1960s. While we were singing his songs, Woody Guthrie, who wrote "This Land Is Your Land" and hundreds of other great songs, who worked with Pete Seeger establishing the modern singer-activist tradition, lay dying of Huntington's chorea not far from where I lived. Huntington's is a devastating neurological disorder, taking the mind along with the body, but all very slowly. Later on, when Arlo Guthrie, Woody's son, started writing and singing songs, the nature of Huntington's as a family disease became clear to a lot of us: would Arlo get sick too? We, the guitar-carrying, tie-dye wearing, antiwar protesting baby boomers, came to have that kind of pseudointimacy you have with famous people. We felt as though they, and the disease, were part of our lives, part of our world.

Huntington's follows the simplest pattern of inheritance. People carry 2 of each of the 100,000 or so genes on the 22 paired chromosomes. One version, or allele, of a particular gene causes Huntington's. If a person carries that allele, and if they live long enough, they are fairly sure to develop Huntington's chorea some time in their lives. Symptoms typically develop in middle age. Each child inherits one of each allele from each parent: so if a parent has the Huntington's version of the gene, the parent will be affected and the chances are 50-50 for each child of inheriting the Huntington's allele and so being affected.

Marjorie Guthrie, who had been married to Woody and had four children with him, all at risk for Huntington's, used his fame to mobilize around the disease. Her organization raised money and raised consciousness, lobbying for yet more research and support money from government, and focusing on service and patient-oriented programs. Many families with Huntington's knew nothing about the disease, did not have even the basic inheritance information you just read, didn't know what was happening—just a family curse, backrooms of mental hospitals, and waiting for the bad blood to turn up somewhere in the current generation.

Woody Guthrie died of the disease in 1967. In 1968 another family began its Huntington's story. Leonore Sabin Wexler was diagnosed with Huntington's. Her ex-husband, Milton Wexler, told their two daughters, Nancy and Alice, that their mother had Huntington's disease, a "progressive, degenerative, neurological illness, that it often caused madness, that it was always fatal, and that both you and your sister each have a fifty-fifty chance of inheriting the illness yourselves." It wasn't, of course, the beginning of the family's Huntington's story—there was an uncle, a grandfather before him. But it was this family's moment of encounter, their start as a Huntington's family.

Milton Wexler had been an attorney, later a psychologist. He was well placed, and did major fund-raising and organizing. Nancy Wexler, just entering graduate school at the time of her mother's diagnosis, became a clinical psychologist whose work focused on Huntington's and other genetic diseases; Alice Wexler wrote a family memoir of their Huntington's story. The Wexlers, as individuals, as a family, and as a research foundation, put their faith in science, and specifically in genetic research. Find the gene: therein, they believed, lies the answer.

The saga of finding the gene is a tale that reads like a great novel, a classic quest, as the family set out to find the cause, the gene, the killer lurking. It is a tale that spans continents, that connects people around the world and that is, ultimately, successful. After a fashion.

One of the problems with doing research on family diseases is that families are, especially these days, often rather small. Even with a 50 percent risk rate, the disease can be elusive. One aunt has two kids, say, and one is affected, but that one gets hit by a car and dies when he is 38, undiagnosed. Another aunt has two kids, both girls, both have the disease. One of those kids has two kids: the daughter has the disease, the son not. Looks like a disease that runs in the women only in that family. With the small sample size that one, two or maybe three children in a typical family give you, it is hard to draw a solid map of the disease pattern.

That is why people call Salt Lake City the "company town" for geneticists. Mormons, whose church is located there, tend to have large families, and as part of their religious practice keep careful genealogical records. But the Mormons didn't seem to produce any particular Huntington's families. So while that has been a useful place for a number of quests, this one went elsewhere. A group was found in Venezuela, poor villagers, relatively isolated, with astonishingly high rates of Huntington's. Rates were so high that some people had two parents with Huntington's, standing a 25 percent chance of inheriting *two* copies of the Huntington's allele—and it takes only one copy for the person to have the disease. Much could be learned from these people.

This area of Venezuela, around Lake Maracaibo, became the scene of an extraordinary research project, descended upon by teams of health workers, scientists, and Nancy Wexler herself, displaying her own biopsy scar as proof of her membership in the community at risk. First, skin biopsies were taken; later blood alone was sufficient to do DNA analysis. And, in the way of DNA research, first a marker was found. Not the gene, but a predictor: people who inherited that marker were likely to get the disease; people without it not. Only later was the genetic error, the actual allele that causes the disease, located.

Locating that marker, finding the spot in the billions of ATCGs that make up the 22 pairs of chromosomes plus the X and Y chromosomes, is no mean feat. As Nancy Wexler described it, "It was as though, without a map of the United States, we had looked for the killer by chance in Red Lodge, Montana, and found the neighborhood where he was living." From that original luck, and the hard work of narrowing the search to the specific gene, it was eventually learned that the Huntington's mutation is a series of repeats, more than 45 CAGs in a row at the start of the gene, with longer repeats indicating an earlier age of onset of symptoms.

So far, so good. We have the makings here of a terrific made-for-TV movie. They hunted, the vibrant, lovely young woman herself at risk, the scientists mobilized, the poor villagers devastated by disease, lined up to give blood and help in the hunt—and they found it. They found the predictor and they found the gene.

But now what? This is where the story stops flat. They knew that with too many repeats of CAG at the start, that gene on the short arm of chromosome 4 causes Huntington's. But they didn't know what the protein is supposed to be doing when it is working right, didn't know how to stop the mutation from causing Huntington's, didn't know how to fix any of it. This is what makes genetics stories less than satisfying as a genre. Even when the quest is successful, what do you have? They didn't have the cure. If the Wexler sisters, either or both, did inherit that Huntington's gene from their mother, they were no less likely to suffer and die of Huntington's after the quest than they were before the quest began. The villagers in Venezuela are just where we found them, still dying of Huntington's.

There is no cure. Some recent research has focused on the biochemical level and looks promising. The cure, if and when it comes, may then have nothing at all to do with the gene. Adding a "good" allele won't help—these people have only one gene that causes Huntington's; the other is good. Can you "snip out" the repeats throughout the cells of the body? No. Can you counter the effect? Possibly, but now we're right back to where we were to start with, needing to understand not the genetics, but the chemistry as it plays out in the nervous system.

So with no cure, all genetic research brought was prediction. Do people want to know that this is their fate? There is general agreement that it would be wrong to test children, that in something so momentous, one should be an adult, able to give or refuse fully informed consent. So does someone, sometime between their 21st birthday and their 40s, when symptoms will probably start to show, want to know this? One of the strongest and most poignant expressions of people's ambivalence is how many people show up for testing only after they have started to show symptoms. The counselors and testers know, just from the time spent in the pretesting interview, that the result will be positive. Is the person seeking a miracle of reassurance, a last-second reprieve? Or a confirmation of an awaited fate met at last?

Because this is not a "carrier" kind of disease, people who want to do prenatal testing, aborting affected fetuses, risk learning their own status: if the fetus doesn't have the Huntington's allele of the gene, then maybe the parent does, and maybe the parent doesn't. Even a parent with the disease has one healthy allele to pass on, and so is working with a two-sided coin. But if the fetus does have the gene, then the parent at risk inevitably knows where it comes from, and learns his or her own fate at the same time that they seek to avoid bearing a child with such a fate. And, because this is a late-onset disease, prenatal diagnosis and selective abortion are particularly problematic for many people: how long does a life have to be to be worth living? How bad does the end have to promise to be to cancel out the value of the 40-odd year start? What is, when you come right down to it, the meaning and purpose of life?

There are lots and lots of other stories like this one. Few have such a famous and appealing cast of characters, few have the drama of that large sad village in Venezuela, but the outlines, the genre, remain: a family curse, traced and found, but not vanquished.

The significant variations in the story are around how horrible the disease is, and how early or late it strikes. Vicious diseases that strike in infancy may make prenatal diagnosis and selective abortion seem to be the answer. Or using donated eggs or sperm and not passing on the disease. Or adopting, or meeting one's nurturing needs in other ways, not parenting at all. When it comes to the more variable diseases that strike early, people who would be parents face even more troubling questions about how much they want to pass on some, but not others, of their genes. They have to do a calculus of how good are the qualities they believe they can pass on in their genes, against how bad are the qualities of the disease they may also pass on. Always the question lurks of how bad a disease has to be to make a life not worth living. For diseases that strike later in life, in adulthood, these questions are still there, along with questions about reading one's own fate along with your potential child's fate— not in your stars but, as James Watson promised, in your genes.

There are so many of these diseases, each story different, all telling the same sad tale. Familial Alzheimer's is a kind of Alzheimer's disease that runs in families, that strikes earlier than the more random-seeming old-age version. There are rare and strange diseases, with long and complicated names, terrifying symptoms winding through families. There is cystic fibrosis, which has been yielding to treatment, not genetic treatment but more traditional medical care, and is becoming more and more a chronic disease that people can live with well into adulthood. And there are the family cancers, most famously now the "breast cancer families," where that too strikes earlier and probably harder than the more usual, random kind of breast cancer. Each of these stories is different, each family has a different quest, its own journey. And each time the same questions come up: What makes a life worth living? And when is a life not worth living? When do we want to know our fate? And when is knowing our fate the worst possible fate?

I'm not going to tell all of these stories, not even brush the surface. I'm going to turn my attention to cancer. Cancer has been the biggest selling point for the Human Genome Project. Science and medicine have been promising to cure cancer "in your lifetime" for much of my lifetime now. I'm certainly ready for them to succeed. But is cancer a "genetic" disease? Can we learn anything from these quests, these gene searches, that will really help with cancer?

Is the genome the ultimate crystal ball, the place where we will learn what fate has in store for us? Is the book of our life written, and sealed, at the moment of our conception? Or are our lives and our cancers the product of our social arrangements, of how we live our lives on this planet, of how we organize our collective and our personal lives? Not long ago we looked to the environment to understand cancer. Now we've switched our attention from power lines to cell lines.

From the Breast

Cancer is many diseases, and it has come to mean many things. I want to explore how thinking genetically changes our thinking about cancer; and perhaps how thinking about cancer changes our thinking about genetics.

Much of what I have to say is about cancer in general, cancer as a process, cancer as an illness. But it is hard to talk about "cancer" in general. The skin cancers that fair-skinned people so commonly have are more of a nuisance than anything else; advanced melanoma is a death sentence; breast cancer is a political and social issue as much as it is a disease; lung cancer is solidly linked to smoking in our minds and in our data. You can't talk about "cancer" and include all of this, everything from childhood leukemia to old men's prostate cancer, and speak coherently.

I'm going to use breast cancer as the touchstone here. Breast cancer has become a current "social issue," a concern that goes beyond the individual. C. Wright Mills distinguished between trouble, what happens to a person, and issue, something that challenges the values of a society. Breast cancer has become such an issue. It's always been trouble: a diagnosis of breast cancer has always been bad news to get. But now breast cancer has gone beyond the people directly involved; it is subject to discussion in newspapers and magazines, included in television stories, discussed in political campaigns.

In point of fact, we are probably making both too much and too little of breast cancer. To even suggest that breast cancer isn't an out-of-control epidemic is positively heretical. Yet some feminist health activists have been saying just that for quite a while now. It was something of a relief to find the same thing being said in the *Atlantic*

Monthly in an article by David Plotkin, a reputable physician and researcher. I copied the article and started mailing it off to all kinds of people who thought I was crazy when I said similar things.

Plotkin wrote of the good news and the bad news about breast cancer—the good news that breast cancer is not the imminent threat, the epidemic that so many women have come to fear; the bad news, equally heretical when feminist health activists have been arguing it, is that medicine has gotten pretty much nowhere in curing it. I want to look at that, look at the data we have on breast cancer, on the push toward early detection, the changes in diagnosis and treatment, but even more, I want to look at the changing image, the changing metaphor that is cancer.

Breast cancer itself is many different diseases. One of those seems to be an inherited disease, a disease that "runs in a family." It is possible to have a "gene for" breast cancer; like all "genes for," it speaks in probabilities. For the small number of women who carry such a gene, the odds of developing cancer during the course of their lives are indeed distressingly high. The first reports were that 85 out of 100 women carrying that gene would develop a breast cancer, and 15 would not. The numbers rapidly dropped, like a bad stock, but with good news for the affected women: the numbers I'm hearing now are around 55 percent. Even at the start of the research, it was known that women had lived into their 80s carrying this gene without ever developing breast cancer. But a lot of women, too many women, with this gene have died young of their breast cancers.

All of the "inherited" cases of breast cancer together do not account for more than 7 percent of the total number of cases of breast cancer. Most breast cancers occur in people who do not carry this gene or any other particular predisposing gene for breast cancer. They don't inherit breast cancer. They just *get* breast cancer.

But do you "get" breast cancer? Do you "catch" it? Does it "strike"? Attack? Develop? Grow? Occur? Our ideas about cancers, all cancers, including breast cancers, have changed as we have come to think genetically. As we think at the level of the cell, what it is to "have" or "get" cancer changes.

Cells reproduce themselves throughout the body; some of our cells continue to divide and replace themselves relatively frequently. All of the cells are descendants of the original zygote, the fertilized egg. From that one undifferentiated cell, all the specific and varied body cells originate. As the embryo develops, a three-layered ball is formed. The epithelium, the cells that form the skin, line the hollow organs and all passages of the respiratory, digestive and genitourinary tracts are all descendants of the outermost embryonic layer. They keep on dividing throughout life, replenishing themselves. Cancerous growths arising from these tissues are called carcinomas, the most common of cancers. Sarcomas are the more rare but often more deadly cancers arising from the descendents of the other embryonic layers.

Cells that divide frequently are the most vulnerable to cancers. Because they are dividing, they can go wrong—mutations most often occur at that moment of division. And that moment of division is a moment of vulnerability, when the carcinogens, the whatever-it-is that causes cancer, can have its effect. Cancer can occur deep within the cell, arising out of its own inborn error, and cancer can occur to the cell, arising from the cell's exposure to carcinogens. It is an interplay, a dance, a flirtation with death and with life, as each cell divides and divides again, divides over and over throughout our lives.

Breast cells continue to divide, to replenish themselves, and so epithelial cells in breasts can give rise to cancers. So too can they in colons, bladders, prostates and lungs, skin and uteri—all epithelial tissues are especially vulnerable to cancer, though cancer can occur almost anywhere.

Thinking of body parts as descendants of common embryonic layers is a strange way of thinking about the body. We have other, more comfortable, more familiar ways of thinking about various parts of our bodies. Parts are private or not, more or less on display, functional, capable of "acting up," sources of pain or pleasure. Different places in the body have different meaning, for the individual and for the culture. Meanings come out of our personal biography—my once-broken left ankle is very different to me than any other joint in my body. But meanings also come out of shared history. Marilyn Yanow recently did a book on the history of the breast. The meanings of the breast have certainly changed over the years, even in our lifetimes. I suppose you could do a history of any body part or organ, but it made sense to do a history of the breast now. It didn't surprise me to see it in a Sunday *Book Review,* part of the everyday popular culture. Madonna and madonnas, wonderbras and silicone, La Leche League, breast pumps, bikinis and topless—breasts aren't what they used to be.

It is cancer of the breast that I will often use as my example, as my touchstone, but it is cancer as an *idea* as well as cancer as an illness that I am interested in. Cancer is an idea that has changed over time, that is changing now. Thinking genetically changes our thinking about cancer, and thinking about cancer changes our thinking about the body.

Cancer as (Not) a Genetic Disease

People are dying of tuberculosis, of gunshots, in car crashes and the occasional plane crash, of infections they picked up in hospitals, of pneumonia, of diabetes, of heart disease, of strokes, of weird viruses, of AIDS of course, of cancer certainly. We're all born in so remarkably similar a condition, and then we die in so many different ways, of so many different things.

Take heart disease: that kills a lot of people. Heart disease has a genetic component, undoubtedly, as probably everything does. There are forms of early-onset heart disease that run in families, that parallel the early-onset familial cancers. Diabetes too. But heart disease, stroke, diabetes—none of these things speak to us the way that cancer does. If you want to understand illness at the millennium, if you want to understand the *body* as we now experience it, look to cancer. That's the disease we are using to write the new body. It is cancer that is telling us who we are, what we are made of, and what will become of us.

It is not, oddly enough, AIDS. There's an irony to AIDS. It is of course the contemporary plague, the disease you have to respond to when you talk about illness at the close of the millennium. It attaches itself to two of the great concerns of America these days: sexuality and drugs. It is a disease, too, of identity politics, attaching itself to "communities" within. It is a disease one wants to compare to tuberculosis, a disease that takes the place of tuberculosis in our imagery. TB was a disease of fashion, its contrasting pallor and flush setting the style, and yet a disease of poverty—and so too with AIDS. It is what the artists die of, and the poor folks. This is a very rich and beautiful story, and if only it has a happy ending....

AIDS is in some ways a quaint disease, an old-fashioned disease. It is the disease of "our" time, but it is also the new TB-like disease, the polio-like disease, writing a new script but with a very old plot line. The anthropologist Emily Martin has studied how American ideas about the body have changed from the 1950s, the era of polio, to the 1990s, the era of AIDS. The earlier model, the model that came with the twentieth century and reached its peak in the 1950s, postwar, polio years, was the disease as invader. Germs. Germs were everywhere, and we protected ourselves with cleanliness. People thought about the body differently then, Martin reminds us: the skin was a barrier, and all contacts between the outside world and the inside, through cuts, breaks, vulnerable membranes, openings of any sort, were dangerous. We built a fort, an antiseptic moat, a shield of cleanliness. Women did that—women were the sentries, the guards, who protected their families by cleaning. Listerine, Lysol, mouthwashes, Band-Aids, and detergents—the smells of my youth. Illness came from outside, and maintaining health meant keeping the illness out, keeping the germs away.

By the 1970s and '80s a new image was emerging, of an actively healthy body, fighting back. Germs got in all the time, there was no hope for that. But the body could, if it were strong and healthy, handle it. The immune system was discovered and was incorporated into the understanding of the body. Getting ill now seemed less a question of what germs came at you, or even what germs got *in*, but how strong, healthy, responsive and flexible an immune system you had.

The irony of AIDS is that it brings us full circle. The germ comes in, and there goes the system. It attacks the very source of health, and it does so in the old-fashioned way of breaching the barriers.

If you can't rely on the immune system to protect you from AIDS, then you have to protect the immune system from infiltration by HIV—and so the barriers come back up. Like the 1950s mother sterilizing everything that came in contact with her baby, boiling and scrubbing a safe swath around the child, the 1990s mother has to protect her child from exposure, keep her family safe and clean. Do you want an HIV-positive baby drooling on your baby's toys? They can keep explaining how delicate that virus really is, and yet . . . We want to snatch our children and ourselves back into the fortress of cleanliness. Polio was spread by the innocent play of our younger children; AIDS by the more subversive play of our older children. But with AIDS, as with polio in its time, danger comes from outside, and safety is believed to lie in the monogamous, heterosexual nuclear family. It is a very 1950s story at heart.

The "genetics" of HIV that people are learning is the takeover story. Like a computer virus, the infiltrator "takes over" the cells, rewrites the commands, and shuts down the system. The "genetics" here is not part of the larger morality play. It is just the mechanism the infiltrator uses. Polio, when you think about it, didn't eat away muscles. It too was a virus that rewrote cell instructions. But that's not the way people thought about it at the time. The story, the plot line, is that something comes in and attacks and destroys. A very old story.

It is cancer, and especially breast cancer, that is writing the new story, that moves us forward into the new thinking, the new geneticism. Breast cancer is the competing disease, the *other* plague of our time. The competition with AIDS comes right to the fore in the ribbon display: in response to the angry, flaming red ribbon of AIDS, breast cancer offers us the pink ribbon. If AIDS is the disease of dirty boys and their innocent victims, breast cancer is the disease of innocence, of mothers, grandmothers, aunts, and sisters. It is a disease that grows at home, not out there in dirty places. It is the disease that both reflects and in turn reinforces the newer, genetic model of disease.

It wasn't always thus: breast cancer was, and not all that long ago, unmentionable. Cancer itself was hushed: the big C, the anonymous "long illness" mentioned in obituaries. Cancer was dirty and it hit dirty places: the breast, cervix, uterus, bladder, kidney, colon, prostate—privates of one sort or another—and places one didn't want to think overmuch about: liver, pancreas, lung—the less pleasant organs, entrails and guts.

But that isn't what cancer is now. Cancer is no longer a disease of organs and guts, a disease of the flesh of the body. Cancer is a disease of the cell, of the program of the body, a genetic disease.

The idea, the philosophy, the ideology of genes is much older than what we are now calling genetics. People have long held the idea that we inherit from family. Our genesis, our beginning, is in our genes: our essence comes from those before us and is passed down to us.

That idea, as I argued earlier, is very specifically rooted in patriarchy: that we are of our fathers, that our fathers make us from their seed, and that we unfold from our fathers' loins while curled in the safety of our nurturing mothers. The patriarchal assumption places our essence in a seed, in the piece of our fathers that turns into ourselves. Nurturance, environment, the world itself can help or hinder, but the essence, the basic limitations of what we can and can never become are written in the seed, engraved in the genesis.

When people think of things as genetic, as in our genes, that is the idea to which they hark back: that we carry within us part of those who preceded us, that aspects of our ancestors have made us, that pieces of them unfold in us. We are no longer a classic patriarchy; we recognize too the seeds of women, the genetics of the egg as well as the sperm. So when we have a grandmother's nose, a grandfather's talent for music, a mother's argumentative style, a father's sense of humor, we see these pieces of others in us, we see history repeating itself, and we say, "It's in our genes."

When geneticists speak about genes, they are speaking of stretches or segments of DNA. Those segments, when in the sperm and the egg, are indeed our "inherited genes." But those segments in a breast cell or a prostate cell or a lung cell are still genes to a geneticist, but not "genes" in that sense of history moved along from parent to child. When geneticists study cancer, they are observing, studying, trying to understand how the DNA segments operate in the cancerous cells. Very occasionally, that has to do with the DNA segments as they were "passed on," in the sense of inheriting a "gene for" cancer. But mostly what they are looking at are the mutations, the changes, that turn cells of individual people from normal cells of the body, ordinary breast, prostate or lung cells, into cancerous cells spreading within the organ and metastasizing throughout the body.

From the perspective of the geneticist, it makes perfect sense to think of cancer as a genetic disease: it is a disease that occurs within the DNA of the cells, and an understanding of the changes within the DNA segments will perhaps provide an understanding of the cancer. But that is *not* the same as thinking of cancer as a "genetic disease" in the old sense that preceded modern genetics—as an inherited, passed-along-the-family disease. Very few cancers show any sign of being "inherited" or "genetic" in that sense.

It is not that we are all so stupid we can't understand that. We can certainly understand the differences between "genetics" as the study of the workings of the DNA within cells and "genetics" as the study of inheritance. The confusion is within the language, but the confusion is also within the presentation.

A number of people are now studying and have written about how "genetics" gets represented in the news. The pattern is pretty much what you would expect: a series of headlines about the discovery of "the gene for" this or that, and then a lot of qualifications, fudging, hemming and hawing about what exactly it all means, and oftentimes a (back-page) retraction at a later date. Those of us who have made this "our issue" read those articles carefully, pounce upon the qualifications, the uncertainties that follow the headline. We read all of it, and read between the lines as well. Just as those who have made Bosnia, or the space shuttle, or welfare reform or school districting their issue can read, and read between the lines of, articles on those topics, understanding richly and fully what is and what is not being said. But for everybody else for whom that is *not* the issue they focus on, we see the headline, glance at the article, get a sense of the issue, and fold the paper as the train reaches our stop—or flip off the radio, zap the TV, close the magazine and move on.

In the genetics articles, the sense has been given, over and over again, that there are "genes for" lots of things, and that cancer is a "genetic" disease. The general sense is that people have, or do not have, the genes for various cancers, and they are *essentially* (in their essence, in their genes) doomed or spared from the start.

A couple of years ago a news article on the radio declared that the gene for colon cancer had been found. I was in the car with a friend for whom this is not a big issue. I heard the news that a gene that accounts for a significant proportion of the 10 percent of colon cancers that are believed to be genetic had been isolated. In other words, they seem to have found a cause of about 10 percent of 10 percent of colon cancers. The remaining 99 percent were not explained by this gene. It was a public radio station, and a thoughtful report, and all of this was said; I heard it. My friend, a bright enough, thoughtful enough person, heard that they'd found the gene for colon cancer. I pushed the conversation with her for a minute, and yes, she realized that they did say they found the gene for the inherited colon cancer. Whatever. Basically, the sense she got was colon cancer was a genetic disease, and they were closing in on the genes for it. Maybe it wasn't as close as she first thought, but that was the direction we were heading in.

It is not a stupid mistake: it is catching the general thrust of all of these stories. The frame for genetics stories is the finding of "the gene for": that is, if you will, the plot line. Then there are some holes to fill in, some finer stuff about what percentage or what variation of the disease, how common the gene is, and how useful this will or will not be for prevention or for cure. But the basic story line is there: they are finding the genes that cause disease, and cancer is a genetic disease.

Understanding Cancer

Cancer is a powerful metaphor, as Susan Sontag once explained. And as times change, as our understanding of cancer changes, as our ideas change, the metaphors change.

We are in a time of a crystallizing, splintering of the self. Cancer is no longer thought of as a rotting agent attacking the self, but as an intrinsic part of the self, programmed, present in our very cells. The battle imagery is still very much there, but this new idea begins to push the battle inside, creating a civil war.

Sontag talked about cancer as a metaphor for unremitting evil. Calling something a cancer was to call it "unqualifiedly and unredeemably wicked," thus justifying an all-out attack, unleashing our full fury against it. Hitler did it, calling the Jews a cancer. That justified radical treatment, destroying the beast and all of its tentacles or claws, excising what needed to be cut away to be sure to get it all.

Cancer, the crab: grabbing hold, eating away, an other, an invader within. Hitler could call the Jews a "cancer" precisely because of the imagery of cancer as alien, other. Cancer cells are not the normal cells of the body, they are cells turned against the body, invading, stealing life away.

Sontag thought that cancer would lose its power as a metaphor for evil when it became mundane: when a single, lowly, in some sense pathetic cause was found, as for tuberculosis. The magic, the mystery, the mythology of the disease, which could be kept alive in fashion, in art, in sensibility, could not be maintained on a bacillus. That little bug-like microscopic thing could not support the mythology of TB, and it dissolved. TB became a disease, just a disease.

Sontag wrote at a moment when viruses were the new frontier in cancer research, when viruses, rather than genes, were blamed for cancers, and immunology seemed the answer. Metaphors of "natural defenses" were gaining ascendence, and doctors wrote books for mass audiences with titles like *The Body Is the Hero,* celebrating the body's own military apparatus, and the potential for the mind to activate that defense.

As the immune system itself gets reduced to the genetic level, the discussion has veered off into oncogenes or "cancer genes" and anti-oncogenes, both the disease and the ability to fight the disease programmed in. While the war is still being fought with the patient as the battlefield, and the destruction produced by the cure is often as vicious as the destruction by the disease, the discussion increasingly moves to the cellular level, and "war" is replaced by "instruction." Cells are to be programmed to recognize cancer cells, to be made smart weapons, destroying only the targeted enemy.

There is a way in which one can talk about these ideologies, these ways of thinking, only as they are in the process of change, just as they are struggling for ascendency or just as they are starting to lose their hold on the mind. Sontag could see cancer as the metaphor it was just as a breath of change was in the air. Genetic thinking offers us a paradigm for understanding cancer, but this perhaps is a moment when we can see that we're going too far with the metaphor of information, programs, computers, and need a better way of thinking about cancer.

One of St. Thomas's proofs of the existence of God is that since all things are naturally still, God is necessary to explain the existence of motion. That's interesting, but are things naturally still? The planets spin, loop and whirl, the whole solar system sails along. A rock is made of atoms, the essence of movement. Nothing is still, ever.

Trying to explain cancer by saying it is growth is like St. Thomas trying to explain motion. Growth and motion are the natural states of affairs. Cells in the body are always growing, dividing, reproducing. The cancer question isn't what makes cells divide; the cancer question is what makes them not stop.

Why don't the cells of the retina just keep on going, cover the entire head? Starting from the single cell, the zygote, how do all the cells become organized, grow where they're supposed to? When you think about it, the real miracle is that we are not all one big undifferentiated, madly growing cancer. The real miracle is that out of all of that constant movement, the appearance and experience of stillness can be created, made to seem natural.

Talking about the body as active at the cellular level is like thinking of movement in rocks. We know it, but it is in some sense counterintuitive. I know about atoms and yet I have an experiential understanding of solidity and stillness. I understand what St. Thomas was driving at, because stillness does feel more natural to me, and it is movement and not its absence that *feels* like it needs explaining.

So it is with the body. It is only relatively recently, after all, that the circulation of blood was discovered, and the experience of the body came to incorporate that understanding. Now when I feel pulsing, I think I feel the blood pushing through. The circulation idea has entered into my experience of my body.

Understanding cancer as today cancer is understood, is to understand the fragmentation of the body, its active and constant motion at the cellular level. It was just in 1860 that the German physician Rudolf Virchow established the cellular theory of the composition of tissues, that the body is actually composed of cells. How much of that understanding really has penetrated? When I see my blood, am I understanding that I am seeing blood cells suspended in liquid? When I feel a muscle strain, am I feeling and understanding that in terms of muscle cells? These ideas begin to enter my mind, and I can easily talk about them, but they are not fully incorporated into my experience of my body. As Emily Martin points out, even just a generation ago, in my own childhood, the body was understood as less differentiated, more of a solid mass inside a protective shield.

It is not just the speed of "scientific progress" that I am thinking about here—it is the speed with which our deep understanding, our folk wisdom, our lived experience, ideology and mythology of the body has changed. I live in a circulating body. I feel and believe that. I am not sure that I live in a cellular body, a DNA-instruction-based body.

Before we understood the workings of cells, cancer could be understood only in terms of its gross pathology, what the body experienced, what the eye could see, what the cancer wrought. Cancer ate away tissue, invaded the body, spread throughout and ultimately destroyed the individual. When a benign growth occurs, a non-cancerous tumor, it is encapsulated. Cancerous growths have no boundaries. A benign growth, as it grows larger, will push against organs and skin; a benign tumor in the breast will, if it keeps growing, make a bulge. A cancer will eventually grow straight through, pushing aside skin cells or eating away at the skin, like a rotting fruit in a paper bag. Cancer was the crab, grabbing hold, eating away, destroying from within.

With a microscope, one looks not only at the ulcerated breasts, the rotted-away lung, the destroyed organs, but at the cells that are cancer and the cells that are not, and the disease becomes defined on the cellular level. That definition goes past the people at the microscopes, and it becomes "our" understanding of cancer—not a crab within, but the cells of our own body gone mad.

Cancer is the disease of the fragmented, cellular body. The cells, in the process of dividing and replicating themselves, rewrite themselves into something else, de-differentiate, re-create themselves as something new, something old and evil.

Because I have truly incorporated the idea of circulation into my experience of the body, when my blood pressure goes up, when my "blood boils," I can feel its pounding surge. I experience my heartbeat, my pulses, as part of that circulating system. It has been made real for me, it was part of how I learned my body, and it is part of the socially mediated construction of my body. My blood flow also makes sense to me in terms of the rest of the physical world in which I live. I have ongoing contact with water pressure, faucets, valves and pumps. I see the hose tense as the water moves through until it pours out; I feel the pressure build when I play with the spray. I've been doing that all of my life, and when my body is explained to me that way, I understand it. I understand "my plumbing" as my mother liked to put it, more in reference to my genito urinary system than my cardiovascular, but still, plumbing is plumbing.

My kids grow up programming VCRs, playing computer games, thinking of "instructions," "bits," and "programs" like I thought of faucets and hoses. It is part of their lived world. When they learn that their body is made up of cells programmed to replicate themselves, that metaphor can be incorporated into their experience in ways it will probably never be incorporated into mine. And when they learn that programs go awry, that errors occur in copying, they will understand that as clearly as I understand strokes and aneurisms.

To understand cancer is to understand the body in its new incarnation, as the product of information, as the object writ by DNA.

Early Detection

My father died of cancer when I was a child. Now that I have a son just a few years younger than my father was when he died, I am tempted to say he died when *he* was a child. He was 28. I was the oldest child: I was eight. I'm the one that takes after my father, always was. The family lore was that I was my father's, Linda was my mother's, and Jeff, the baby, was a mix. I had the "chocolate pudding eyes" he had, the wide feet, the dark hair, the brown-as-a-berry (by white folks' standards) summertime skin. The year after he died I got glasses too, just like Daddy, not like anybody else in the family.

I have been waiting for my cancer all my life. The most noteworthy part about middle age for me is that it is too late to die tragically young. Relatively young is still possible, but shockingly young is over. I expect this cancer to come, and with each cancer scare (the lump-in-the-breast rite of passage for middle-aged women, the unexpected bleeding, the more subtle stuff that cancer-phobes note) I go right past the "denial" stage and straight to burial. I make plans. If I'm optimistic, I make plans about treatment options: Maybe, if it is the kind of cancer against which I might have a small shot of buying a few years, maybe then I'll try chemotherapy. I plan the party to shave my head—why wait for the hair to fall out? I'll pierce my ears—never have, always felt too barbaric. Barbaric sounds right, suddenly. But mostly I start grieving for my younger children, the ones I will shortly be abandoning. No denial, no bargaining: the other shoe has dropped.

At the last cancer scare (and they're not that often, don't get me wrong, just every few years) it was a lump-in-the-breast. One friend who loves me well but will never understand me started a pep talk about fighting it. Another, who knows me better, knew what I was really feeling: relief. I was as much relieved that the wait was over as anything else. And *that* scared me.

My father's cancer is not, objectively, a threat to me. But I didn't understand that, had no reason to, until I was fully adult: too late for it to penetrate all the layers that have been built up, too late to build up all the layers that have been penetrated.

He had an undescended testicle. In the era when he was born, nothing was done about it. Now it is known that left up in the body cavity, a testicle can develop cancer. That's what happened to my father: it had spread through his body before anyone knew what was happening. It was years after he died that the information about what really caused his death came to the family. New research or whatever. When my son was born, where others might count toes, I counted balls. Both present and accounted for: he could be named after my father.

Counting balls and seeking safety, trying for security in signs and portents—that is not so different from checking genetic markers.

What signs and portents do you use? An annual physical? Pap smears? Mammography? Chest x-rays? How do you know you're healthy? How do you really *know?* Maybe you feel fine—but so what? What does that prove?

Being healthy isn't as obvious as it once was. Many of us now have internalized a new notion of "seeming" versus "real" health. Seeming health is the way you feel and all the standard measures of healthiness. You can feel fine, think clearly, have good muscle tone, a steady heartbeat, good appetite, rosy cheeks. But that no longer means you are really healthy: all kinds of death and destruction can lurk within, the hidden diseases, the silent killers.

Screening programs sell us this idea: don't trust your feelings of health. Get tested. Blood-pressure testing is offered to me as I walk down the street. Toll-free numbers are advertised for lupus. The American Diabetes Association sends me mail, urging me to see my doctor and be tested. And mammography is offered on my campus.

These are all examples of presymptomatic screening programs: they look for diseases that may exist in the person but have not yet made themselves felt. It is an interesting concept actually: you can have the disease, but not have any sign of it. Diabetes then becomes not the disease that causes the problems associated with diabetes, but the condition of having high glucose levels. You are what you test. A person with high glucose levels has diabetes, *is* diabetic. The test changes people's status—the way they think about themselves, the way that insurance companies think about them.

The value of presymptomatic screening is that some of the symptoms, the problems caused by a disease process can be avoided if the disease is treated early on. If high blood pressure causes strokes, for example, then finding the high blood pressure before a stroke, and lowering it to avoid the stroke, makes sense. High blood pressure is by definition a condition that could exist only as a result of testing; it is not a "disease" on its own. You have to have a test of blood pressure before you can have a "condition" known as "high blood pressure." But given the test, you can have the condition and be treated for it.

Every screening program has a debate behind it: How accurate is the screening test? How many people are missed who have the condition (false negatives), and how many are picked up who do not have the condition (false positives)? More complicated questions frame the debate as well: Is there a treatment and how effective is it in avoiding the development of symptoms? How much risk does the treatment entail, a risk imposed on people who might never have developed the symptoms anyway? Would it cause irreversible damage to just wait for the appearance of symptoms? And balancing these risks, who should be screened—for whom are the risks of the screening and the treatment greater than the risks of just waiting? It's a judgment call, which screening programs are worth it, and for whom.

Screening programs are very popular. Part of that is the result of marketing. There is money to be made in screening people. The larger the population screened (Ashkenazi Jews, people over 50, all women, all newborns) the more money there is to be made. So screening programs are often aggressively sold.

Screening programs are also popular with health-care providers, physicians and other practitioners. There's not much medicine has to offer healthy people; screening programs provide a service. They structure patient-physician interaction to some extent: blood is drawn, urine sampled, blood pressure taken.

And screening programs are popular with their consumers as well. They allow you to feel like you are doing something, being responsible for your health. And there's even more to it. An older woman I know is exhausted, taking care of a husband with Alzheimer's. They see a social worker for counseling. You have to take care of yourself, too, the social worker told her: you have to do something for your own needs, not just take care of him. So the woman scheduled a mammogram. It's a way of taking care of herself, of being good to herself.

Screening programs never guarantee health of course: everybody has some story of someone who got a clean bill of health one day only to be sick the next, someone who dropped dead the day after a physical. But they do, inevitably, offer a kind of reassurance, lay to rest some particular worry for a while.

Cancer screening has had varying success—as is to be expected with such a varied collection of cancers. Cervical cancer screening, with a Pap smear, represents the oldest cancer screening, named after George Papanicolaou, who developed it in the 1930s. Cervical cancer screening became, in some ways, the model for cancer screening. But Pap smears are not prototypical of cancer screening just as cervical cancer doesn't represent all cancers.

Cervical cancer is like skin cancer: the cervix and the skin are epithelial tissue, which is more or less readily accessible. The Pap smear is a simple scrape of surface cells off the cervix, placing them onto a slide where they can be viewed under a microscope. Pap smears also pick up some cases of endometrial cancer—the endometrium is the lining of the uterus. Cervical cancer is a slow-growing cancer, which spreads mostly "geographically," to the regions it touches rather than by metastasizing to distant sites. A Pap smear can show normal cells; cells affected by some disease or infection; displasia, which is a change in cells that may develop into a cancer; a cancer *in situ,* that is, a localized, noninvasive cancerous growth; or a fully developed cancer. In the language of diagnosis, these are class I through V results. Cervical cancers are easily treated: the surface of the cervix can be more thoroughly scraped or cut clean for a class III or IV diagnosis, or the cervix can be removed for a developed cancer. The test itself is safe, involving no radiation, no chemicals, no biopsy. Pap smears can miss cancers, but because cervical cancers are slow growing, routine testing will still pick up cancers before they have spread, even if a single test misses it.

Yet even cervical cancer screening, the very model of cancer screening, has been shown to be less effective than it ought to be, and for a reason that speaks to the ironies of screening. Women who have been screened and told they are cancer-free seem to ignore symptoms that would otherwise have sent them to a physician. Postmenopausal bleeding should be a worrisome sign: it might indicate a cancerous growth. But women who have been screened have learned exactly, and too well, what screening programs teach: don't trust your body and your sense of health, trust the test. Trusting the test that told them that they were okay, they then ignored the signs that other, untested women see as matters of concern. The end result is that cervical cancer screening does not bring down cancer deaths nearly as much as it should, in theory. The more the cancer in question varies from the slow-growing, easily accessible model, the less effective the screening programs will be.

Breast cancer screening has been the subject of much controversy. On the one hand, early detection is practically a religion: it is a widely and deeply held belief that with early (enough) detection, breast cancer is curable. On the other hand, the hard data to support such a belief simply doesn't exist.

One of the complications is the way that we think of "cure." We don't actually speak of cancer cures; we speak of cancer survivals. If a woman has a cancer that will absolutely, positively kill her on the day of her 70th birthday, no matter what (to take a hypothetical example), and you find that cancer when it has spread and become symptomatic, she may have only months to live after her diagnosis. If you find it earlier, she may have years. If you find it very early, she may pass the five, even the ten-year survival rate. She'll be just as dead at 70, but with earlier diagnosis you can increase the number of years she spends as a cancer patient, and so the number of years she "survives" her cancer. In that sense, early diagnosis inevitably increases survival—assuming that nothing in the treatment itself kills her any earlier.

Breast cancer rates are rising. That may be a fact; or it may be an artifact of increased screening. With each new wave of screening, diagnosis goes up. As doctors have more often and more carefully checked for breast cancers, they have found more. As women have learned to do breast self-exams, they have found more. As mammography has become routine, it has found more. More and more cancers have been found. But the death rate for breast cancer has remained surprisingly stable in the 60 years that records have been kept. In 1935, 26.2 women per 100,000 died of breast cancer; and in 1992, 26.2 women per 100,000 died. One possibility is that there are more breast cancer cases, but they are harmless cancers, cancers that the women would outlive even without diagnosis. They may be the result of "lifestyle" changes, longer years of regular menstruation, unbroken by pregnancies, lactation or malnutrition. We enter puberty earlier and menopause later, with fewer pregnancies in between. If that causes a new, less deadly kind of cancer, as David Plotkin suggested in the *Atlantic* article, then "For all we know, the chief effect of mammography has been to disguise our inability to cure the old cancer by burying it in cases of new cancer." Or there may not even really be all that many cases of "new cancer," just new diagnoses, more testing finding more cancers.

A lot of faith has been placed in mammography, in moving diagnosis back to an earlier stage of tumor development. But tumors grow by cell doubling, the cells dividing and dividing. Mammography typically diagnoses a tumor that has doubled almost 27 times;

another 8 or 9 doublings and the tumor can be felt by hand. "To argue that earlier diagnosis provides an important benefit, one must believe that the tumor is considerably likelier to spread in the eight or nine later doublings than it was in the preceding twenty-seven." Unlikely.

Earlier diagnosis might not work, but it continues to give the appearance of working: women do survive longer after an earlier diagnosis, after all.

More profoundly, I think, it is the logic of early diagnosis, or presymptomatic diagnosis, that opens the path for genetic testing. Presymptomatic testing seeks to find an identifiable disease or disease process before it does damage. Genetic screening seeks to find something that isn't there yet, seeks to read the future. If it is better to diagnose a disease process before it produces symptoms, then isn't it better to diagnose it even before it appears, before it has even begun to exist? Genetic screening and testing for "predispositions" or "tendencies" builds on early or presymptomatic diagnosis, moving the diagnosis back yet again, to a point preceding its physical embodiment, and into the realm of codes, of predestination, of tragic flaws within. With genetic testing for the "breast cancer gene" a three-year-old girl can have a diagnosis of breast cancer before she even has breasts.

In a sense, this is a return to an earlier understanding of disease as intrinsically related to the person who has it, not to some outside force that causes it. You have cancer not because of something you did, or something that happened to you, but because of who you are. This places the disease deep into the body, internally, intrinsically, essentially within.

Cancer in this way of thinking becomes an inevitable playing out of one's essential nature. The cancer is the fatal flaw, not just an invisible crack in a beautiful vase, a crack we can find if we look carefully, even before it opens up to the naked eye. No, the cancer moves into the very nature of the being, a flaw built into the very clay of which the vase is made. One doesn't "get" cancer in this model. One's cancer emerges.

Breast Cancer

Breast cancer is not a "genetic disease." It is not there, born in, waiting to come forth.

You wouldn't know that if you read the newspapers. The widely publicized "breast cancer genes," BRCA1 on chromosome 17 and BRCA2 on chromosome 13 together account for maybe 5 to 7 percent of breast cancers, and 5 to 10 percent of ovarian cancers. Those two genes together seem to account for 90 percent of all of the inherited cases of breast cancer, all the breast cancers that "run in families." Other genes, not yet found, may account for the other 10 percent of these inherited breast cancers, but all the inherited breast cancers together still leave 93 to 95 percent of breast cancers unexplained.

When the genes were first found, it was claimed that women who had one of those genes had an 85 percent chance of developing breast cancer. The genes were found by studying families that had a high rate of breast cancer. Once the research went beyond those few families, the number dropped rapidly. It is altogether possible that those original family members shared other risks—other genes, environmental factors. Now it seems that only about half of the women with BRCA1 or 2 develop breast cancer in the course of their lives.

It turns out, then, that a very small number of women carry a gene that dramatically increases their chances of developing breast cancer.

Much has been made of this gene in the media: I've read, heard, been earnestly told that they've found the gene for breast cancer. Women who have had one grandmother develop breast cancer in her old age tell me they're "at risk" for breast cancer, it "runs in the family." A seventy-year-old woman, aunt of one of my graduate students, was told by her family doctor to have both of her healthy breasts removed because her sister developed breast cancer. If it weren't so tragic, if people weren't so terrified, you could say this was getting silly.

People are coming to believe that breast cancer is a genetic disease, a family disease, that some of us are doomed and some of us are safe.

Not so. Between 93 and 95 percent of all breast cancers are not familial, are not "genetic" in that sense. And almost half the women who have one of these identified "breast cancer genes" (and maybe even more than that once the numbers are all in) won't get breast cancer.

What does happen to women who learn they have BRCA1 or 2? What exactly are they supposed to do about it? Finding these genes has opened up a troubling can of worms for those affected women. While breast cancers usually develop later in life, the breast cancers associated with BRCA1 and 2 often occur in women in their 30s and 40s. Mammography is not very effective in finding cancers in younger women: their denser breasts make the growths harder to see. Besides, more frequent mammography starting at a younger age would mean more radiation exposure—not necessarily a smart thing to do to people who are at increased risk of cancer. So early detection, the first thing that pops into mind for a high-risk group, may not be a good thing and may not work all that well.

What is rather glibly called "prophylactic mastectomy," removing the breasts while they still seem healthy, is sometimes suggested. One doctor suggested it to a young woman I know, calling it "preventative medicine." "Preventative medicine," she replied, "is wearing a hat in the rain, not cutting off your breasts." There isn't even data to show that such extreme measures will work: no one knows how early in breast development the breast cancer cells start, and they may very well have seeded throughout the body before surgery. And no one knows how slight an amount of breast tissue left behind might be enough to develop a cancer. So even women ready to have both of their breasts removed in their youth cannot guarantee escaping breast cancer.

American women face an added burden, in that they also risk losing their health insurance coverage, or being made unemployable because uninsurable, if word of their BRCA1 or 2 status leaks out.

Some families want prenatal diagnosis, aborting affected daughters. A woman who's lost a mother, an aunt, and a sister to breast cancer while they were still young might seriously consider selective abortion to avoid passing on what feels like a family curse to her daughter. Does that make any sense? We're back to basic questions about how long a life has to be to be worth living, what the meaning of a life is, and who is to judge. Should my father have been aborted had his cancer death been predictable and not preventable? Not for my sake, certainly, but consider my grandmother—she was a bitter, sad old woman who lost both of her children to cancer, my father in his 20s, my aunt to lung cancer in her 50s. If those cancers had been "familial" cancers, what should my grandmother have done? And who could *know?* Who could judge?

Families react to this diagnosis of BRCA1 or 2 in their families in about the same ways that families react to all of the hard stuff: some better than others, some with support, some with estrangement between the "affected" and "nonaffected" members. Guilt, grief, anger, sorrow—about what you'd expect. But all of this has nothing to do with the vast majority of women who have breast cancer. This has nothing to do with the "breast cancer epidemic," an epidemic of *awareness* even more than of disease.

In the days before the increased awareness of breast cancer, an older woman with a small lump in her breast, a lump she may or may not have been aware of, might have become sick and died of the cancer as it spread elsewhere in her body. What kind of medical attention she was receiving, what her symptoms were, how old she was—all of these things will have affected how much attention was paid to tracking down the exact cause of her death, that original breast cancer. Slow-growing cancers may have escaped all attention, allowing the woman to die of a stroke, heart attack, car accident, or any of the diseases one becomes vulnerable to in advanced age. It is still the case that some old women die with small breast lumps that may or may not be in the process of spreading cancer through their bodies. They die of something else, the cancer undiscovered. That scenario becomes less and less likely as the aggressive push for diagnosis widens the net, catching more and more breast cancers.

This spiraling process—increased awareness, thus increased reporting, thus more increased awareness—is a common one, not just for breast cancer or for diseases in general, but for any number of life events. It is one of the ways "social problems" are created: not only out of an increased number of such events—chain snatchings, abused children, car jackings, whatever—but out of an increased awareness. The more we know about these things, the more we see them, the more we recognize them, report them, seek out information about them.

Oftentimes the events that are reported in the media or in our daily talk, recorded officially and informally, turned into "issues," are events that can be turned into little morality plays. They offer a message, a story from which we can learn. Breast cancer has lent itself to a number of such morality tales, turned to advantage by a variety of interests. The first is the one I started with: that breast cancer is a genetic disease, and that work on mapping the human genome and on genetic research will somehow bring us an end to the breast cancer epidemic.

But there are a lot of other morality tales written on the breast cancer theme: some breast cancer stories are part of the "backlash" phenomenon. Breast cancer is presented as the "price women pay" for progress, for giving up the "natural" state of affairs. By not having so many children, by not having them so early, by not nursing them so long, we are told we make ourselves vulnerable to breast cancer. In the ugliest version of this story, and on the flimsiest of data, abortion is claimed to "cause" breast cancer. In a sense then, breast cancer serves us right.

More often breast cancer is used as a more general morality play about progress: thanks to the wonders of modern medicine, we are told, we live longer and so die of diseases of older years, like breast cancer. The apparent impotence of modern medicine to accomplish very much with breast cancer is then turned around into a paean to the success of modern medicine: the very success which then permits us to live long enough to have breast cancer. Like all of the cancers and diseases of later years, breast cancer becomes, in this story, the next hill to take in the long and so far successful battle against disease.

But breast cancer is also used to tell a very different story: breast cancer as a "feminist issue." We hear about the barbaric treatment of women, the cavalier attitude of surgeons who lop off breasts, the insensitivity to women displayed by medicine. All of which I think is probably true, but actually tells us more about medical management of cancer than it tells us about medical misogyny. Not that I am one to doubt the misogyny of medicine as an institution or as practice: it is that in this case, ideas about cancer are even more powerful than ideas about women.

Breast cancer as a disease is as ancient as can be; breast cancer as an issue dates back to the early 1970s. It was a time of sexual revolution, when suddenly modesty about breasts seemed absurd. Topless bathing suits appeared, and going braless was both a fashion and a political statement. Cancer itself began to be more public, both a more acceptable way to die and a more possible way not to: cure and survival seemed increasingly available. Two prominent political wives had public bouts of breast cancer: Betty Ford and Happy Rockefeller did for breast cancer in their day what Rock Hudson did for AIDS in his. Without dying. And it was a time of the women's movement, when women, women's issues, women's choices and rights and interests moved into the public eye.

At that time, the management of breast cancer was a nightmare. American surgeons combined biopsy and mastectomy into a single, two-stage procedure. The woman was admitted to the hospital, brought into an operating room and anesthetized. The lump was removed, the biopsy stage of the procedure, and prepared for frozen section. That permitted an instant (if arguably less accurate) diagnosis. If the diagnosis was cancer, the surgeon moved directly to mastectomy. The woman had no opportunity to participate in any decisions, or even to have a chance to adjust to what would happen to her. The reality of it was thrust upon her as she awoke. From the patient's point of view, it meant going into the operating room without knowing if she would wake up within minutes with both breasts, or hours later with one. Most women, about 80 percent, had a negative biopsy, and came through with both breasts.

The surgery itself was a far more mutilating procedure than is done today. It was developed in the late 1800s by an American surgeon named William Halsted, and was known as the Halsted radical mastectomy. The surgeon removed not only the breast, but the pectoral muscles that underlie the breast, leaving skin-covered ribs. Also removed were the lymph vessels and nodes extending into the underarm. It was a deforming surgery and a debilitating one, the arm on that side often likely to swell and be extremely painful.

These then were the two major complaints of the women's health movement of the time: that the surgery was unnecessarily extensive and that the one-step procedure denied women any decisions about their own care. Given the context of the rest of the women's health movement, which found both the overtreatment of women and their silencing rampant, these seemed like apt feminist concerns.

But if you look at the management of breast cancer in the context of the general management of cancer, neither of these complaints are unique to breast cancer or to women. Rather, they were indicative of the medical approach to cancer.

Cancers grow unencapsulated—they trail off here and there, a few cells extending into what looks like normal tissue. Understanding the general process of metastasis, the "seeding" or starting up again of small cancer colonies from an original cancer site, surgeons cut wide, hoping to "get it all." Halsted claimed to bring the local recurrence rate from the 58—85 percent it had been, down to 6 percent. He believed that extensions into the pectoral muscle were common. Current concerns have shifted from local recurrence and spread to metastasis throughout the body. The kinds of breast cancers surgeons see these days are different. The particular tumor described in Halsted's first public account of his surgery was the size of a "hen's egg." The woman had first noticed it when it was the size of a "pea" a full year earlier. With today's greater cancer awareness, such a large tumor is very, very rare. If that had been a fast-spreading cancer, the woman would have been dead of spread to bone, brain, and the rest of her body before it reached egg size.

Fear of local spread also explained the combining of biopsy and mastectomy. Surgeons were afraid to biopsy and close up, possibly leaving a cut-into cancer inside. They feared that the cancer cells would be seeded by the scalpel itself. In breast cancer, that fear turns out to be simply ungrounded. Biopsies do not spread cancer. Lumps are now removed under local anesthesia, the diagnosis is made, and then women have time to make decisions about further treatment.

There is no question that the earlier management of breast cancer diagnosis and treatment was a dreadful experience, and one shared by many women, not just those who had breast cancer, but all those who went through this biopsy process. It was easy to see it as part of the patronizing, dehumanizing, belittling treatment of women by medicine.

But an interesting and telling parallel can be found in the medical treatment of testicular cancer. While physiologically breasts and testes aren't parallel, there are social similarities. Having a testicle removed, like having a breast removed, is a private mutilation, hidden by ordinary dress. And one can function sexually with one testicle or breast as well as with two, but both surgeries do hit at vulnerable sexual identity. When testicular tumors were found, the patient was operated on under general anesthesia. The scrotal sac was opened, and if, to the surgeon's eye, the tumor was suspected of being malignant, the testicle was removed. Not biopsied, removed. The fear of spreading malignant cells by cutting into a cancerous testicle was so great that it was felt to be better to risk removing an occasional benign testicle. Just think what would have been said had surgeons been women.

Testicular cancer is not the only parallel to be considered. There is also the contemporary debate on cancer of the prostate. The U.S. has taken an aggressive stance toward prostate cancer, pushing for early detection and treatment—much like breast cancer approaches. In this instance, the early detection being pushed is a blood test, the PSA (Prostate Specific Antigen). The test was heavily marketed, including the use of a public relations firm, a "Prostate Cancer Awareness Week," and celebrity patients like General Norman Schwartzkopf telling men they "owe it to their families" to have the test.

Prostate cancer, even more than breast cancer, appears to be a part of aging. Autopsy studies of men who have died of causes unrelated to cancer show that the incidence of prostate cancer increases throughout the life span, becoming virtually universal by the time men reach their 80s. Finding and treating every single case of early prostate cancer would probably do considerably more harm than good: that is what the debate is about. The surgery risks making men incontinent and impotent, and it has a one percent postoperative death rate. Some men do die of prostate cancer, but it is quite possible that they would die at the same rates anyway: the death rate of treated men appears to be the same as that of untreated men. The Prostate Cancer Intervention Versus Observation trial, begun in 1995, was an attempt to definitively resolve the issue, by doing a solid controlled clinical study—but by February 1997 only a few hundred of the required 1,050 men were enrolled. Most men refused to participate in a study that might assign them to "watchful waiting." Similarly, the 1985 U.S. clinical study showing that a simple removal of a cancerous breast lump had the same outcome, the same survival rates, as mastectomy, required the participation of Canadian surgeons—the American ones too often would not participate, unwilling to risk less aggressive treatment.

So I find myself in the very bizarre position (as anyone who has read my previous work on obstetricians and gynecologists can attest) of defending American medicine against charges of misogyny. It is not that medicine is not misogynist; it is that it is irrelevant in this instance. The treatment of breast cancer has been about cancer, not about breasts and not about women.

Still, women's responses to the treatment of breast cancer certainly said something about women's attitudes toward medicine. Women were at the forefront of a consumerist health movement that has profoundly affected the ways doctors deal with their patients.

Just 20 years or so ago, enormous secrecy surrounded cancer. Doctors lied all the time to their patients, and particularly to their women patients, protecting them from the awful truths about their own bodies. It seems like a million years ago now. They don't lie about death any more. They are blunt—perhaps to a fault, but honesty and bluntness now rule. There are still enormous problems. Doctors too often do not deal well with patients: people who are carefully selected for their ability to do well on standardized admission tests do not necessarily do well in interpersonal skills. And people who are trained to win do not handle failure well—and for medicine, death means failure, and cancer too often means death. So the horror stories are still there, the awful things doctors say to their patients, the dreadful ways cancer patients are treated. But now the stories are about bluntness, and facts rather than silences are used for fences.

The honesty, though, has permitted more patient interaction through support groups of various forms. In breast cancer this took flight in a dozen different directions: environmental groups, groups seeking cancer families, sexuality concerns, alternative treatments, spirituality groups. Breast cancer, perhaps more than any other cancer, broke the silence.

A lot has changed about cancer in the past twenty years. Susan Sontag wrote that cancer was a secret, a shame, cancer deaths ugly, unromantifiable. She wrote just as Elizabeth Kubler-Ross was beginning to be heard and the hospice movement was growing. We have gone from being a society that was death-denying, silenced and silencing about death, to one where *How We Die* made it to the bestseller list, describing in crushing detail just how death happens. Cancer is an ideal-type death, in the Weberian sense, a classic, prototypical death. The cancer death captures the imagination in ways that heart disease, diabetes, stroke do not. Only AIDS has similarly gripped the imagination: the wasting, the pain, the suffering. These are fully orchestrated deaths. Cancer offers us heroic battles, an enemy within, a disease that captures the imagination along with the body. And the full-scale warfare that medicine wages against cancer has been more warlike than many of its other battles. To fight cancer, medicine does take on the body, sacrificing breasts, limbs and organs like battalions, losing battles in hope of winning the war, destroying the land in hope of liberating it. "Fighting the good fight" and "surrendering with good grace" have both become common ways of thinking about cancer.

The changing nature of woman as metaphor, of cancer as metaphor and of the breast as metaphor all conspired to make breast cancer a very different kind of disease than it had been. Public women had public breast cancers and breast cancer could never be the same again. Some women won, and their beautiful, Amazonian, single-breasted bodies came to stand for something other than mutilation. I remember a magazine photo from the late 1970s of a woman's nude torso, a long-stemmed rose tattooed across a mastectomy scar. Other women died, but they did not die in ignorance, protected from knowledge of their own bodies. They went as a new kind of feminist hero, creating new deaths, new rituals and rites, constructing community and sisterhood out of breast cancer.

But now I am afraid that breast cancer has become, oddly, a kind of "post-feminist" issue, a "safe" women's issue. Abortion, pay equity, motherhood concerns, domestic violence, sexuality—all have been divisive for women. Our multifaceted responses have shown that we are not one woman, one voice, one essential being. But breast cancer is safe territory: we are all against breast cancer. It worries me to see breast cancer used in these many narratives: antifeminist, feminist, postfeminist. These stories are constructed to reinforce our womanliness, our essential sameness to each other, our difference from men.

The breast cancer story focuses our attention on one of the very few ways women die differently than do men. We commemorate breast cancer with pink ribbons, turning it into a women's solidarity statement. But without the feminist context, what's a "women's issue"? Of course we're "against breast cancer," but all of the trappings of a political movement—marches, demonstrations, ribbons—won't turn it into a political issue. Politics has to do with power, with social justice. Sontag was right: Breast cancer is a *disease*. And when placed in a genetics frame, when it is argued against all reason, data, and logic, that somehow breast cancer is a genetic disease, a gene some women have and some women don't, whither solidarity?

Contradictions

Rich or poor, cancer strikes. And rich or poor, people die of cancer. You can't just buy your way out of it. Rich people, the children of rich people, famous people, people with all of the money, love, help, and support in the world still die of cancer. It threatens us all. That is a truth, and a good one to remember: there are forces more powerful than money in this world.

It really is a good message, an important one. We are told that money is the root of all evil, and that money can't buy happiness: but we know, each of us, that there is a lot of happiness in our lives that money could indeed buy. So it is a good thing to remember that health is more precious than money, that life itself is a gift beyond dollars, and all those other pious and true things we say.

But that is a lesson, I find, that we may have taken a bit too much to heart. When I teach social epidemiology, the ways that disease is distributed within a society, my students find it very hard to hear me, hard to believe me. I tell them, and I show them, and I graph and chart and prove that the richer you are the more likely you are to be healthy and to have healthy children; and they—not all, but so many, over and over again—mark that "false" in a true/false question, argue against it in essays. Rich people, they tell me, get sick; rich kids get leukemia; it happens to everyone. And of course they are right. It does happen. What changes with social class, social position, is not the possibility, but the probabilities. Your odds shift.

In the earlier eugenics model, that was understood and expected: poor people were believed to be of less sturdy stock, they were expected to be sickly. Poverty of body, mind, spirit and pocket were all expected to go together in the world. The microeugenics logic of our time severs that connection: illness, we believe, occurs at the level of the individual body, not the social body.

Americans are good at seeing the individual. We think individualistically, we think psychologically and not sociologically. We can see the person; in sophisticated, nuanced ways we can understand individuals, their motives, concerns, actions, behaviors. What we are less good at, as a people, is seeing a people: seeing the whole, seeing the social, the system within which individuals move.

Diseases occur within the bodies of individual people; but diseases are also characteristics of populations. And as individual experiences, they are largely unpredictable. We can make only the vaguest of predictions: I probably will get a cold and not a brain tumor this year. Even with the best and most accurate of predictive techniques, even with the most powerful genetic predictive tests, the predictions are always presented as "odds," as probabilities, as chances. Not as facts.

At the population level, the story is different. We can, with a startlingly high level of certainty, predict rates of disease. Name any disease you care to, breast cancer, strep throat, tuberculosis, athlete's foot, colds, brain tumors, whatever: I can find you a chart showing the cases over the past years, and a solid predictor for next year. We can predict these diseases, and we can be astonishingly specific about how many people with each of the many different diseases there will be.

This is always a difficult idea to hold in our heads: that at one and the same time something is so random, so unpredictable, so capricious, and at the same time, so predictable, so routine, so ordinary. This is one of the ongoing tensions in medical care: the shocked person getting the diagnosis; and the doctor who would be shocked if *no one* got that diagnosis this year. The diagnosis is part of the everyday work-world for the doctor—predictable, dependable, a certainty. And it may feel like the end of all that was everyday, predictable and dependable for the patient.

One kind of understanding of cancer is to understand it at the cellular level, to follow its story within the body of the patient, to make it understandable at the level of the individual. The other understanding is to follow the pattern, the distribution, the social picture.

When I was a kid, I had the idea that ambulance crews ought to carry small cans of red paint, and whenever they picked someone up who'd been hit by a car, they should splash the paint in that place. It would warn the next person crossing that it was a potentially dangerous place. And if several splashes overlapped each other, another added before the last had worn off, it would be a clear and immediate demonstration of a particularly dangerous spot, and maybe someone could do something about that. I've seen variations of that idea carried out now and again—white crosses by the side of the road in some countries—and I must say, I still think it is one of the better ideas I ever had in my life. It makes public a private tragedy that may very well have public roots: if it is just one fatal moment, there will be one splash. Kids dart out, brakes fail, minds wander. If it is structural, systemic, caused at least in part by forces outside of the individuals involved—a badly sighted corner, a poor arrangement of stop signs, a badly designed drive—then we will all know about it, and maybe we can fix it.

That was, I think, the underlying idea behind the red ribbon for AIDS. If all of the people who feel touched by AIDS wear those ribbons, the public, shared, collective nature of this tragedy will be increasingly obvious. And that is the notion in turn of the pink ribbon, the breast cancer ribbon.

And yet. The change we seek to make with all of this public awareness does not seem to be a public change. We may be asking for more public money—but it pours into individualizing explanations and responses: Get genetic testing. Eat less meat. Have more mammograms. The structural explanations and the structural implications are glossed over.

My mother found two (two!) two-year-olds wandering the streets in the past week. Just loose, diapers bulging, one naked otherwise, the other in early-morning pyjamas, wandering around unattended. The weather has gotten warm, doors have been opened, outside has some appeal for the first time in living memory of a two-year-old, and, in each case, they seem to have just slipped away. It happens. One was being minded by a baby-sitter, and it is tempting to get angry, to lay blame on inattentiveness, to speak sternly to the young, surprised woman who opened the door to find my mother standing there, holding the child she hadn't even known was missing. The other child had wandered off from around the corner, and it was 20 minutes later before a frantic woman finally spotted my mother, again standing there holding a two-year-old. The mother's knees buckled, she sank down into the grass in front of my mother's place, couldn't even hold herself upright as she reclaimed her child. Either kid, two feet tall and not terrifically aware of the dangers, could have been hit by a car. And in either case, there would have been enough blame and anger to go around to last lifetimes, generations.

Those would have been purely private tragedies. And yet we have designed homes, cars, streets, a way of life, that make such a tragedy pretty much inevitable. We've designed our lives so that cars drive up to our front doors. We've designed our front doors to open right onto the streets. We've designed the cars so that things low to the ground are difficult to spot. We've arranged the care of young kids so that you have to have eyes in the back of your head, constant vigilance by one often overworked individual, or the kid will occasionally slip away, wander off, get into one sort of trouble or another. It is a private tragedy when a child is hit by a car; and it is a structured, political, group decision about our priorities. If we don't make some massive changes in the way we live (and drive) I can guarantee you that some two-year-olds will be hit by cars next year, and can make a pretty close estimate as to how many, what kinds of neighborhoods they live in, what kinds of traffic patterns they face. It is random and it is predictable. It is private and it is public. It is personal and it is political.

And so it is with illness, and so it is with cancer.

We do not all stand equally placed with regard to our likelihood of getting any particular disease. Yes, some of the difference lies in our genes, our inherited tendencies that are passed on within families. But much of the difference lies in where we are placed within the society. The patterns of diseases, distributions of illness within a society, are just as real, structured and measurable as patterns at the level of cells.

The simplest social pattern to see when looking at cancer rates is the geographic distribution. U.S. cancer rates are highest in the northeastern states of New York, New Jersey, Connecticut, Massachusetts and Rhode Island. They are lowest in Utah, Wyoming and Idaho. The differences aren't small, either: they account for about one-third of the cancer deaths in the high-risk states. Think about that: if you live in the Northeast and have lost local family and friends to cancer, then approximately one-third of those deaths were preventable. They wouldn't have happened had all of you been living in Idaho.

How can we think about that? The language of causality doesn't seem appropriate. Can we say that living in New Jersey *causes* cancer? Hard to say, when most of the people who died of cancer did not live in New Jersey, and most of the people who did live in New Jersey did not die of cancer. What we can say is that living in a northeastern state increases your individual risk of dying of cancer: the rates for the population vary by state, and risk is the other side of rate.

If any given individual moves from New York to Wyoming, they presumably lower their chances, or at least their children's chances, of developing cancer. But if we all moved from New York to Wyoming, we don't know what would happen. We would probably turn Wyoming into New York in so many ways, the population density, the pollution, the changes in the environment, that it might not work as a cancer prevention at all.

When individuals' DNA predisposes them, when they have "a gene for" cancer, their chances of getting cancer increase, just as when a person lives in a state with a high rate of cancer, their chances of getting cancer increase. We've become comfortable with saying the gene "causes" the cancer, but less so with saying the geography causes the cancer.

For geographic patterns we have at least a map in our heads, a way of understanding and thinking about "living in New York" or "living in Utah." For social patterns, it is so much harder, especially given our national propensity to make believe that social class doesn't really exist.

Social class—if we think about it at all—is an abstraction for most Americans. Lots of commentators have pointed out how naive Americans are about social class. We just about all of us check off "middle class" when asked to place ourselves in the class system. We understand some people are rich and some are poor, but tend to see that as tiny, fringe exceptions to a vast middle-class center. That has never been true, and gets further and further from the truth as the century comes to an end, pushing more and more of us out of the "middle." But the imagery is still there, the notion that we are all middle class.

The social-class portrait of cancer is complicated. Most cancers are inversely related to social class, just as most other diseases are: the richer you are, the healthier you are. More money, less disease. Two exceptions to that pattern are breast cancer and prostate cancer. It shouldn't surprise us then that breast and prostate cancer are both described in terms of "epidemics" and "threats," while colon cancer, bladder cancer and lung cancer seem somehow to have taken a back seat.

Some, and maybe all, of that difference that goes against the standard class relationship is a reporting difference. Poorer people are less likely to have the presymptomatic testing that uncovers breast and prostate cancers at an early stage. Since some breast cancers and probably most prostate cancers never become lethal, a failure to diagnose means an escape, for all practical purposes, from the illness. An undiagnosed, nonlethal breast or prostate cancer might just as well not exist.

But even where there are genuine social-class differences in occurrence rates—as there may be with breast cancer (rising with social class) and surely are with lung and bladder cancer (falling with increasing social class) the relationship is tangled. You can't say

being poor *causes* bladder cancer, or having money *causes* breast cancer—any more than one could say living in New Jersey causes cancer. The differences are always in rates, which go up and down for
various segments of the population. Slice it in different ways—class,
sex, race/ethnicity, occupation, percentage body fat, lifestyle differences—and different cancer portraits or "maps" can be drawn.

If you flip back to the individual level, to the cellular level, something like "occupation" or "social class" doesn't exist. One is hard put
even to understand "sex" or "race/ethnicity" at the level of a lung
cell. At the cellular level, there is the DNA inside the cell, and the external "environment" of the body, and the world outside that body
as it impinges on what goes on inside. That is, when we think about
cancer, there are the cells "oncogenes" and "anti-oncogenes," that are
thought to constitute the genetic basis for tumor growth and tumor
suppression. Some of these "tendences" and "predispositions" and
"susceptibilities," or more positively "resistences," are "genetic" in the
sense of inherited, running in families. Others are the product of
mutations—accidents that have happened in the course of the history of that cell. The cell has, at any given moment, its own internal
life, with its own history.

Impinging on that cell are the stimulants, the carcinogens, the
cancer-causers, and the cancer-protectors, that come from outside of
that cell. In the broadest possible sense, they are environmental.
Some are the environment of the body itself—estrogens produced
by the body that stimulate a given cell, for example. Some are environmental in the sense of coming from outside of the cell, outside of
the body, like the viruses that stimulate some cancers. Some are from
the environment of that particular body—smoke inhaled, coal tars,
workplace chemicals, estrogens that come from particular meats
that are eaten by that particular person. And some are "environmental" in the sense that we more commonly use the word, like radiation
levels, pollutants that come from water, air and food that enter our
bodies, penetrate our cells, stimulate our cancers.

Cancer, at the level of the cell, is a process. You can't "explain it"
by the presence of a carcinogenic stimulus—not everyone so ex-

posed develops the cancer. And you can't "explain it" by the presence of predisposing mutations in the cell that started the cancer growing—not everyone with such "cancer genes" develops the cancer. Cancer occurs at the level of a cell. And cancer also occurs at the level of an atomic blast. It is caused by changes in the DNA. It is caused by industrialization. It is caused by oncogenes. It is caused by cigarettes.

Cancer represents the tension between thinking socially and thinking individually, between an "environmental" ideology and a "genetic" ideology. Both ways of thinking are with us, and both frame cancer for us. On the one hand, there has been an increased awareness of cancer as being environmentally caused: questions about fat in the diet, about smoking and other lifestyle issues, as well as larger questions about industrial pollutants, power lines and the ozone layer. On the other hand, there is this newer image, this genetic thinking which asks us to think of the cancers as lurking in the cells of the individual. Both ways of thinking draw maps for us, shape our imagination as we think about cancer. Neither map precludes the other: environmental stimulants can goose predisposed cells into action, or can bring about the predisposition.

Even for the diseases of concern during earlier periods, susceptibility varied, of course. Not every child exposed to polio got polio; not every person exposed to TB got TB. Some susceptibility follows the social map: children are differently placed, and malnutrition, harsh living conditions, different patterns of exposure to different diseases, all shape the child's response. And some response was more individual: I'm now coming across data that suggest a susceptibility to polio might lie on chromosome 19. The explanation for the diseases lies in the external "cause," the bacillus for TB and the virus for polio, but there is also an explanation to be had in individual response. In the best of all possible worlds, all lines of research would be followed.

But we never have lived and do not now live in the best of all possible worlds. TB itself is rising again as the environmental stresses that made for vulnerability return, and as the bacillus becomes immune to the treatments. It begins to look like poverty *does* cause TB. As to cancer, just a few years ago the bulk of cancer money was going into viral research: the answer to everything seemed to keep coming up "virus." Now it goes into genetic research. The social epidemiology of cancer, the role of industrial capitalism, gets glossed over as cause moves deeper and deeper inside the individual.

The TB bacillus and the polio virus did not have the financial backing of big investors, of government support, the advantages of public relations firms behind them. The viruses that appear to stimulate some cancers are similarly politically unprotected. Other kinds of environmental stimulants do have backing. Tobacco is probably the easiest to see, as the tide begins to turn, and because tobacco appears to be such an individual choice. Tobacco is still a legitimate business in the United States, its advertising and sales are still legitimate business expenses. Joe Camel was at least a recognizable if formidable enemy.

Whatever, whoever it is that destroys the ozone layer, or lets chemicals slip into my water, air and food is more diffuse. I'm oftentimes not sure just who I should be angry at. The sun—the very sunlight, source of enormous joy to me—is an increasing threat. The public health approach has been largely aimed at *me*, at individuals to "just say no" to the sun too, along with all the other things that are bad for us. There are of course activist, political approaches to cancer that look *up*, blame the destroyers of the ozone layer, the polluters of the air and the water. But mostly we seem to be in an era of looking down—telling individuals to eat more wisely, slather on the skin protectors, wear hats, stop smoking, and go live someplace else if you don't like the air in your neighborhood.

A friend in a public health master's program was doing a paper for a course. She had to develop a program to encourage teenagers to cover up from the sun. This is what they *teach* in a Public Health program? What if the public health movement had, back in its earlier incarnation, directed its efforts not at improving the water supply, but at getting individuals to boil their water, to "just say no" to unboiled water? Perrier and Evian are not public health measures.

But it is worse than that, this new thinking. It's not only about encouraging individual change (and boiling water was an immediate and necessary solution). What if the earlier generation of public health workers had spent their research money on figururing out which arm of which chromosome holds the "gene for" susceptibility to cholera? Do we want them to have figured out which people were most susceptible and what was wrong with them that made them so vulnerable? I have no idea what individual, cellular, internal factors account for variations in susceptibility to cholera. And I don't care. I opt for a safe water supply.

And so, I believe, it can be with cancer. Of course individual susceptibility varies. And for a few relatively rare cases—the small percentage of breast or colon cancers that "run in families"—that susceptibility is so great that it does need to be considered. But for most of us some susceptibility is there, a given. Expose us to carcinogens and we will, as a population, as a people, as a community, develop more cases of cancer. Remove those dangers, and we will, as a population, as a people, as a community, be healthier.

To switch our attention from power lines to cell lines in the search for the causes of cancer, to keep looking down instead of up, is dangerous. To individualize the disease obscures the role of the social world in causing cancer.

Cancer is the product of mutations. But it is also the product of society. We need more than one map if we are to imagine solutions.

Part III

IMAGINING THE FUTURE: THE MICROEUGENICS OF PROCREATION

The Gates of Life

Genetic thinking is a way of understanding the world, but genetic practice is, above all, a way of imagining the future. Understanding for its own sake is not much valued in American culture. We're a practical people. We want progress.

Medicine holds out hope for progress, hope for a better, disease-free world. In genetic thinking, the line between prevention and cure is easily blurred: to cure a genetic disease means to rewrite the code and so prevent the disease process. And the possibility opens up, when rewriting the code, of preventing a disease not just now, of this moment in this person, but forevermore. The new, improved code, the disease-free code, can enter the gene pool and replace disease with health for generations to come.

Because genes are the new frontier in medicine, a lot of attention is being given these days to gene therapy. Gene therapy is treatment that uses genes, stretches of DNA, as therapeutic tools, as hydrotherapy uses water, chemotherapy uses chemicals, and massage therapy uses massage.

It is hard to gauge just how well gene therapy is going to work, when all is said and done. So far, not a single person has been definitively helped by gene therapy, even though more than two thousand have been treated, and they were selected for gene therapy precisely because their conditions looked like they might be amenable to gene therapy. But the optimism is undaunted, and who knows? Remember the Wright Brothers and all that. Pretty much every report I've read of every and any attempt at gene therapy says two things: on the one hand, it's not working. On the other hand, it's just *bound* to work. Ruth Hubbard, who as a biologist reads the more scientific end of this literature than I do, says the reports do not all say it is bound to work. So maybe it's the optimism of the science, or maybe it's the optimism of the media that present it. Sometimes medicine has been spectacularly successful (Goodbye smallpox! So long polio!) And sometimes not (Goodbye thalidomide! So long silicone implants!).

But I'm not writing about genetics as a technology, or even genetics as a science. I'm writing about genetics as a way of thinking. So my question is not How well does it work? but What does gene therapy have to do with how we think, particularly how we think about ourselves?

Here again it is very important to distinguish the geneticists' use of "gene" from the common English meaning of "gene." Of course, we have lots of nongene therapies for lots of genetic conditions: glasses, braces, "nose jobs," cardiac surgery on newborns. Some, maybe a great deal, of what is "wrong" with us, that sends us to medical care, is at least in part genetic, and so far none of the most effective treatments have been genetic. On the flip side, we may soon have gene therapies for diseases that are not genetic.

Say a person gets hit by a car, or has some other kind of accident, and trauma occurs. The spine or the brain is damaged. That is not by any stretch of the imagination what we think of as a genetic disease. Because of the injury the person has a break in the spinal cord, and loses the use of her legs. Or she has brain damage and can't speak or can't remember, or any of a million other dreadful things that can happen to someone after a trauma to the nervous system.

The reason the person is hurt is because the car hit her. That is not "genetic." Yet at the cellular level the reason she remains hurt is because her nerve cells are unable to divide and repair themselves. That inability to divide *is* genetic. Cells do not replicate themselves in the nervous system the way they do in the skin. A big gouge in the skin, and cells divide and replace the lost ones and you're back in business. A similar chunk out of the nervous system, and you're out of luck.

The concept of gene therapy is to change the genes: in this case to stimulate these cells to divide again. Nerve cells are able to replicate themselves during embryonic development, which is how the nervous system developed in the first place. Somewhere in their DNA presumably lies a stretch that could initiate the production of more neurons. Genetic research may find a way to activate that stretch, and that would be gene therapy for traumatic injury.

Gene therapy can also be targeted to the effects of a genetic disease, without touching the genes that produced those effects. There are any number of central nervous system disorders, from trauma to inborn errors of metabolism, that might lend themselves to gene therapy. The treatments may address the disease process, without dealing with the genetic components that cause the disease. Certain substances are needed in the brain, and brain cells typically produce them. If they don't—because of inborn error, trauma, poison, or any other reason—a form of therapy that induces cells, by changing their DNA, to produce the needed substance would be genetic therapy. In other words, if a particular allele of a particular gene on a particular chromosome in a particular zygote causes a person who has grown from that zygote in 2 years (Tay-Sachs) or in 40 years (Huntington's) or in 70 years (Parkinson's) to be lacking some essential substance or overproducing some other substance in the brain, gene therapy might be able to correct that.

Take Parkinson's disease: it results from the death of neurons, brain cells, in a particular part of the brain. That loss can be caused by any number of different things, including certain inborn predispositions. Those neurons secrete the neurotransmitter dopamine. Transplanting human fetal cells that produce dopamine into the affected portion of the brains of people with Parkinson's disease has worked, but the cells die off. Research is being done to make genetic modifications to the fetal cells before they are transplanted, to prevent their dying. Here gene therapy is applied to the cells to be transplanted, not to the individual receiving them. The cure has nothing to do with whatever genetic predisposition the person with Parkinson's may have had that caused the original cells to die.

Similarly, gene therapy for cancer wouldn't necessarily or even often involve cancers that are "genetic," and doesn't address the genetic "causes" of the cancer. There are a number of different approaches to gene therapy for cancer. Some try to make the cancer cells recognizable by the immune system as foreign. Some try to make the cancer cells kill themselves. Some try to cut off the blood supply to cancerous tumors. These are all gene therapies because they involve altering genes. But they are not altering genes that cause cancer, even if the cancer was caused (largely or in part) by a gene.

"Gene therapy" is not necessarily what we think it is. We tend to think of it as changing the genes of the individuals being treated, rewriting their "program." It might very well involve that, but it might very well not. And such rewriting might be correcting genetic error, to make a gene "normal," or it might be, as in the case of getting neural cells to regenerate, changing the cells from normal to abnormal, making them do something they would not normally do in a healthy adult.

So research in gene therapy proceeds, with attempts to find "vectors," mechanisms for getting the repair genes where they would be useful. Genes are put into various kinds of viruses, into little fatty bubbles called liposomes, or just "naked" into the body. This is a big technical accomplishment, and a bigger one if they get it to work. Yet it is still medicine as we know it. The body is missing something and the doctors try to add it. The body has something extra, and the doctors try to remove it. In some larger philosophical sense, it builds on the traditional Western notion of the body at war against disease, with the doctor adding reinforcements on the side of good against evil, whether the new troops consist of aspirin, cough medicine, L-dopa, or a stretch of DNA. The story stays the same.

That is *not* the kind of "gene therapy" people are talking about when they worry about "playing God." We're used to the idea of helping sick people get well, preventing dying people from dying, "playing God" in the sense of taking control over life and death. But we've made our peace with that, almost all of us, and think it is a good thing to prevent untimely death and suffering.

What seems to worry people now is a kind of gene therapy that directly influences future generations. Of course everything influences future generations. Give an antibiotic to a baby with a raging bacterial infection, and you may have changed the course of the future for untold generations. All medicine in that sense can have consequences for future generations, for controlling our evolution.

A few people do worry about that. Some religious groups reject some of the offerings of modern medicine. And, coming from a very different place, some eugenicist thinkers continue to worry about keeping "genetic weaklings" alive to reproduce. But most of us and our society, our institutions, our collective imagination, have come to accept positioning medicine at the exit gates of life to redirect traffic. Maybe God or nature or whatever powers that be intended for some baby to die, but if we have a vial of medicine that will cure that child, we use it. Maybe my father was meant to die young, but believe me, if those who loved him had known how to save him, they most assuredly would have, and they wouldn't have apologized for playing God, rewriting the future or any of that.

What has many people concerned now is the repositioning of medicine not only at the exit gates of life but also at the entrance portals. Genetic technologies begin to make promises not only about who among us will die or be saved from dying, but who, not yet among us, will be called forth. We are starting to ask what kinds of people we will permit to be born, what kinds of people we will create. Using genetic knowledge, genetic predictions, genetic technologies to decide not who will die, but who will be born—that *is* a new story.

That is the story of prenatal diagnosis and selective abortion: genetic testing developed to prevent the births of babies whose lives were believed to be tragedies best avoided. The first such category diagnosed was fetuses with Down syndrome, but other disabilities were also understood to make life not worth living. Prenatal diagnosis was developed as a way of diagnosing fetuses so that the births of those with severe disabilities could be prevented.

Thirty to forty years ago, when prenatal diagnostic testing was being developed, it was standard American practice when a "damaged" or "defective" baby was born to knock out the mother with anesthesia. Lots of women were anesthetized or just made drowsy for birth in that era, but even wide-awake mothers were made unconscious when there was a serious problem with the baby. The drug was used, not for the health of the mother or the baby, but just because doctors feared "hysterical" mothers in the delivery room.

The woman was then informed later, without seeing her baby. It was an open secret that some babies were allowed to die of exposure and hunger in the back of American hospital nurseries, and that survivors with severe disabilities were warehoused in institutional settings, their families encouraged to forget them and have other children. The birth of a disabled child was understood to be an unqualified tragedy from which women should be spared.

It was in this context that prenatal diagnosis and selective abortion were developed, as a way of avoiding tragedies. There are no genetic therapies for genetic conditions diagnosed prenatally: all a woman can do with the information she gets is decide whether or not to continue the pregnancy. Abortion became legal in the United States and made prenatal diagnosis practical: there was something one could do with the information. And prenatal diagnosis helped legitimize abortion: abortions when the fetus would be born with severe disabilities are still among the most socially acceptable of abortions.

The early testing was terribly crude. The genetics involved simply counting chromosomes: they could identify the trisomies, the genetic conditions associated with third chromosomes in what should have been a pair. Some trisomies are incompatible with life. Down syndrome, a third 21st chromosome, sometimes causes the death of the fetus, baby or young child, but a person with Down syndrome can also have a long life. What is best understood, and most feared in America about Down syndrome, is that it is associated with mental retardation.

Not all prenatal testing was "genetic." A variety of diagnostic technologies were introduced, including removing and testing samples of amniotic fluid, the water the fetus floats in, and ultrasound imaging, "seeing" the image of the fetus with sound waves. Some of the conditions so diagnosed were "genetic," like Down syndrome; some were not, notably neural tube defects, which may or may not have a genetic component. These are openings in the neural tube, ranging from small openings at the base of the spine, which can cause minor motor impairments, through anencephaly, a failure of the brain and top of the head to develop, causing a very early death.

But whether we are testing for a genetic condition, a condition resulting from a virus, a condition resulting from an environmental exposure or some combination of causes; and whether the testing is done with ultrasound imaging, with DNA testing on fetal cells in amniotic fluid or mother's blood, with chemical testing on amniotic fluid tapped from the mother's belly: these are all technical questions. Whatever caused the problem, and whatever technology found it, a woman learns that her fetus, if carried to term, would be a baby with disabilities. The fundamental moral questions that are being raised are, when you come down to it, *the* fundamental moral questions. We are asking what makes a life worth living, under what conditions is a life not worth living, and who is to judge.

The earliest, crudest testing—counting chromosomes, looking for gaping holes in the spine—raised all the fundamental questions. Fancier technology, more elaborate and more specific readings of markers and of genes themselves won't change the questions, and won't provide the answers.

Spoiling the Pregnancy

It is no easy thing to be positioned at the gates of life, selecting who may enter and who may not. How serious, how unutterably dreadful must the condition of a fetus be that no life at all would be preferable to the life it would have if allowed to be born? And—always— there is the question of who will judge.

The opening up of abortion "on demand" placed individual women in the position of gatekeeper. Abortion as a right became abortion as a responsibility: an obligation a mother might owe her potential child, sparing it from its own life. It gave women enormous responsibility, but without any real power to affect the world in which they were held responsible. And that may, when you think about it, be the essential dilemma of motherhood.

It would be nice to think that there are some rules we could rely on in these decisions. It would be reassuring to believe that there is a right and a wrong decision to be made. For a very small percentage of Americans, opposed to all abortions at all times under all conditions, there is a right and a wrong. But the way that the problem is generally presented and understood, by the physicians, genetic counselors and bioethicists, is that there is indeed a right and a wrong, solid shores on which to stand, but unfortunately a whole lot of rather murky grey area between them. When I listen to the women who are actually making these decisions, the most solid of the shorelines crumbles, and it is all the grey, soggy mess between.

The ethicists who study prenatal diagnosis, and the medical practitioners who provide it, are often most comfortable with those situations in which the fetus is diagnosed with an inevitably fatal condition: it might not survive the pregnancy, or even if brought to term and born alive, would die shortly thereafter. In those cases, prenatal diagnosis and selective abortion is generally understood to present no ethical dilemmas. An abortion simply brings the inevitable to a more rapid conclusion.

That makes perfect sense, given the assumption that the point of pregnancy, the reason for pregnancy, is to get a healthy baby delivered. If you cannot get that done, then why bother continuing? And that is, in a nutshell, much of the rationale for prenatal diagnosis, and is certainly the logic used for testing for conditions that are incompatible with life. If the baby is going to die anyway, there is no point in continuing the pregnancy: it is a waste of time. And if the woman could have known that and did not, then she has in some sense been duped, made a fool of, wasted her time.

This is the logic that underlies the medical "management" of pregnancy. It is product-oriented, and it is centered on the fetus. The purpose of pregnancy is to make a (healthy) baby, and the point of all prenatal and childbirth management is to achieve that goal.

That sounds reasonable enough. But what about mothers? Mothers want healthy babies, good "fetal outcomes," but is that all we want? I've been involved with midwives and childbirth activists for more than 20 years. What they've consistently pointed out is that the medical approach to birth overlooks the experience of mothers. "Fetal outcome" is generally the only dependent variable that counts: how the mother feels about her body, her husband or partner, her family, her child, her sexuality, herself—all escape measurement, except as they might affect fetal outcome. Doctors will casually talk about delivery "from above" vs. delivery "from below," contrasting a caesarean section—major abdominal surgery with weeks of recovery time—with a normal vaginal birth. Ever since improvements in nutrition took care of rickets, and doctors learned to wash and stopped the infections that raged through maternity wards (two of the chief causes of maternal mortality), the focus of obstetrical care has been on fetal outcome, with the woman variously known as the carrier, host, environment or barrier.

Midwifery, in contrast, is focused on women. In English the word "midwife" derives from the Old English for "with woman." Of course midwifery care tries to help the woman have the healthiest possible baby. But it also means trying to give her a "good birth," a pregnancy and birth that make her feel good about herself as a mother, as a woman. It is not just the making of babies, but the making of *mothers*, that midwives see as the miracle of birth. Helping the woman is what midwifery is all about: and that might very well mean, in a situation where the death of a baby is inevitable, helping a woman and her family come to terms with that in the best possible way.

In the United States, as in many parts of the world, doctors took over childbirth, driving the midwives out. It is only in recent years that midwifery has reinvented itself; it still struggles to establish itself as an independent profession. So when I wanted to get a fresh, nonmedical look at how else we might think about genetic testing, and learn how midwives think about prenatal diagnosis and the explosion of genetics information in pregnancy, I went to the Netherlands. Dutch midwives have been in practice for, I guess, forever. Doctors never took over routine birth, but worked with the midwives. Almost a third of Dutch births still occur at home, with far healthier mothers and babies than our high-tech system has been able to deliver.

Midwives cannot speak *for* the women they attend, but they are well placed to think about how prenatal testing changes pregnancy. I've learned an enormous amount about prenatal diagnosis and selective abortion from interviewing the women going through it. But the one thing you cannot expect from these women is perspective, distance. Midwives see a lot of women over a lot of years. As a community, they have a collective history, a culture of childbirth to draw upon. As one midwife in the small village of Harderwijk told me, dating the known history of midwifery in Harderwijk to 1554 to a midwife named Bella Henderinks van Arnhem, "Don't we build up any authority in all those centuries?"

I met with midwives in small groups all over the Netherlands. We talked about the kinds of prenatal testing that were being done, and what it meant for the women. When we talked about these cases where the baby was going to die anyway, where nothing you could do would give the mother a living child, the midwives asked a question that simply makes no sense in the medical understanding of pregnancy. If the baby is going to die anyway, the midwives asked, "Why spoil the pregnancy?"

Why indeed. I've been working with midwives for so long that the question made perfect sense to me. I didn't even realize that it was a strange thing to say until I shared it with American friends and colleagues who weren't midwives.

You can't spoil something if it has no intrinsic worth. If pregnancy is only about making healthy babies, and the baby is going to die, then the pregnancy *is* spoiled. But that's not the way the midwives saw it. In one group, considering the possibility of a bad outcome, a midwife said, "Well do they have to know it? Let them first have an untroubled pregnancy." And in another, when that idea was being talked about, a midwife leaned over to me, the American, the outsider, touched my hand, looked in my eyes, and explained: "Some of us find a good pregnancy very important whatever the outcome."

I see two ways of understanding their valuing of a good pregnancy under these circumstances of an inevitable bad outcome. One begs the question of what is the point of life altogether. If life is about accomplishing things, then pregnancies resulting in dead babies are pointless. If life is about living, though, if it is just *there*, and we have only a finite time to live, then days spent in joyous anticipation are good days, and days spent in grief are bad days, and prenatal diagnosis of conditions that inevitably cause death simply moves days from the good to the bad side of the ledger for the woman.

The other way of understanding this—and they are not mutually exclusive—is to postulate that pregnancy itself has a meaning and a value in a woman's life, and that for women who want to become mothers, a good pregnancy and a good birth are good things to have.

Consider the following:

> In our practice was a child with a disorder that was not compatible with life. It didn't have a *midrife* [diaphragm]. Intestines up, heart in the wrong place. This woman had a good pregnancy, a difficult delivery, but she looks back on it very positively. The child lived a couple of hours. Of course they are sad about the child, but also had very positive feelings toward the child. And I saw a couple of pregnant women talking about it, and they said, "You could have seen it on an echo" [an ultrasound], which is true. "They should have done an echo, then they could have know." And that is how other pregnant people talk about it, like it's nice to know in advance that something is wrong. Theoretically. But they did not know this woman. This woman is very satisfied that she did not know anything in advance because an echo wouldn't have changed it. Yes, she would probably have had a hospital birth and three thousand echoes and pressure, and now it is at least a nice pregnancy and positive experience of her delivery. The outcome would have been the same in both cases.

The outcome to which she refers is the loss of a child, a loss that was inevitable whether by abortion in the first half of the pregnancy or the death of a born child. Not all abortions are felt as the loss of a child. But for a woman planning on having a child, a diagnosis of a fatal condition has to mean the loss of that planned-on child. She was pregnant, planning on a baby. There will be no live baby, no living child. The outcome, however achieved, is the same.

Because the midwives are thinking about the needs of the mother, they see things so differently. This exchange took place in a discussion of a baby with severe heart disease:

> First midwife: The children's doctor looked at the baby, everything okay. At night, the baby's temperature fell down. It had no chance to live outside the uterus. If I had made an ultrasound the disablement was shown. What will be the profit for this woman? She was pregnant, very happy, had a very good delivery, was very happy. However the baby died 24 hours later. But it died in her arms. What if we saw it on the echo?
>
> Second midwife: No profit, only much worse I think. This was human.
>
> Third midwife: Once I made an ultrasound, the baby was anencephalic, the child had no head. During the pregnancy the woman said goodbye to the baby. That was very important to that woman, so what is profit or loss?

In the medical model, that is readily answered: profit is time saved, loss is time wasted. Rather than waiting for an anencephalic baby to be born and to die, an abortion is an efficient solution. But between the lateness of the diagnosis, and the fact that these are wanted pregnancies, these abortions are not comparable to ordinary, early abortions to get "unpregnant," where the pregnancy itself was a mistake. These abortions, for the woman, are the death of her baby, without the saving grace of a good birth and a good death.

The midwives are considering the process and not just the product. Let me clarify the distinction with a mundane example. Suppose you have a video camera and are filming your kids. They are mugging for the camera, singing a song, playing, laughing and having fun together, when you notice that you have run out of tape. If what you are trying to do is make a tape, then there is no point in continuing—you should tell the kids and maybe try again another day when you have fresh tape. On the other hand, maybe you should just continue to do what you were doing, let the kids finish up, let the fun go on, and forget about getting it down on tape. Partly this is a difference between product and process, but it is also a way of thinking about what the product is: when filming the family, you are also constructing the family, making those very ties between your children that you seek to capture on tape.

Pregnancy is about making a baby, but it is also about making a family, making relationships, making the woman a mother. Even if the pregnancy is not successful in producing a baby, it may very well be successful in its other products. Death and grief and sadness also make a family. By nurturing the woman, her relationship with her partner and her family and friends, her feelings about herself and her lost child, midwives can construct success, satisfaction, *family* even out of death. In this approach, since nothing is going to make the baby any better, one has no reason to learn ahead of time and spoil the pregnancy, burden the woman with untimely grief and a ghastly birth-unto-death.

The case of a fetus whose condition is incompatible with life is as *simple* as prenatal testing gets. Whether you start from the medical model, our standard American notion that the whole point of pregnancy is to make a healthy baby, and if you can't do that there is no point in continuing; or you start where these midwives do, that something good can come out of even these pregnancies, this is still the clearest case.

From the beginning, advances in genetic testing made things more complicated, not less. Tay-Sachs was one of the first genetic conditions for which prenatal testing and selective abortion were made available. At first, the testing wasn't "genetic" testing; it was biochemical testing for a genetic condition. But that too is a technical consideration: find the chemical product that marks the disease; find the marker on the chromosome; or find and decode the gene itself, the upshot is the same. Some woman is told her baby is going to die. Anencephalic babies live for a few days. Tay-Sachs babies live for a few years. Children with cystic fibrosis lived a decade or two, longer now with better treatment; some of the familial cancers come in people's 30s; Huntington's disease comes at midlife. There is no point, we say, in continuing the pregnancy if the baby is going to die right away. How about soon? How soon? How long does a life have to be for there to be a point to it?

Get a Map

At least with the questions about the length of life we know what the measure is. With questions about the quality of life even that is unclear.

Medical practice has been to present Down syndrome as a clear diagnosis. While there is a substantial bow in the direction of the right-to-life movement, and therefore a discussion of prenatal diagnosis just to "be prepared," essentially the testing is for the purpose of selective abortion. There is no way to "be prepared" when given a diagnosis of Down syndrome precisely because the diagnosis doesn't tell you much about what the person might be. In other words, the diagnosis may be clear—see that third 21st chromosome and you can be sure the fetus will have Down syndrome. But the prognosis—what will become of that fetus—is not at all clear. Women contemplating prenatal diagnosis often say that they would terminate the pregnancy for severe mental retardation, but not for mild retardation. What you cannot know from looking at the chromosomes is how severe the retardation will be.

Some ambiguity is built into almost any diagnosis. Some of that ambiguity is the "probability" factor that accompanies any genetic information. But it is more than that. Even if you know exactly what the condition of the fetus is, it is still hard to know what the life of the person might be. I met a lawyer once whose son had a neural tube defect. They knew when he was born that he would most likely use a wheelchair all his life. She said she and her husband were okay with that, came to terms with it pretty easily. She said they talked about it, figured they were both intellectuals and not athletes, and this just wasn't going to make that much of a difference. When I spoke to her, her son was 17. He was a national wheelchair basketball champion. They lived their lives around game schedules.

Books and articles on genetics and genetic counseling used to give you this list of the genetic diseases. The lists got longer and longer. I'm not sure if they even bother any more. Presumably, with 100,000 or so genes, there are easily 100,000 or so places for things to go wrong. Then there are all those diseases and conditions that have "genetic components," the predispositions, tendencies, increased likelihoods. We obviously cannot terminate pregnancies for all of them or nobody would ever be able to give birth to anyone. So what do we do with all these predictions?

After my time working in the Netherlands, a Dutch student of sociology arrived to study with me in New York for a few months. It was helpful to keep seeing America through her eyes, to get a fresh take on things. I got to work one morning and found a clipping on my desk from a *New York Times* report of a speech President Clinton had given. She and the other Dutch student she was rooming with had been debating if Clinton had been speaking ironically or not:

> The Human Gene Project at the National Institutes of Health, also being supported in universities all across America, will one day in the not-too-distant future enable every set of parents that has a little baby to get a map of the genetic structure of their child. So if their child has a predisposition to a certain kind of problem, or even to heart disease or stroke in the early 40's, they will be able to plan that child's life, that child's upbringing, to minimize the possibility of the child developing that illness or that predisposition, to organize the diet plan, the exercise plan, the medical treatment, that would enable untold numbers of people to have far more full lives than would have been the case before.

I assured her that American presidents do not "do" irony in election years.

But it is a bit hard to take it seriously.

A statement like this implies a level of control that boggles the imagination. Surely even presidents do not have that level of control over their children. Try to imagine it: you get your prenatal or new-born printout, your map. You've got the whole range of then-knowable genetic information. Your kid or kid-to-be has, say, increased risk of bladder cancer, double the national average. Half the risk of lung cancer. One-third again the risk of diabetes. Stroke unlikely. Alzheimer's a good possibility. Not to mention good guesses on alcoholism, schizophrenic or manic-depressive tendencies, and whatever else they find—or think they find—on that map.

Now what? That bladder cancer risk: look at a geographic map for bladder cancer. Are you in a high-risk area of the country? Should you perhaps move? If you decide not to uproot your family and your child does develop bladder cancer, how are you going to feel about that? And where should you move to? Presumably you can get printouts like this on each member of your family. They're not all going to look alike.

And those other tendencies: Has your kid with the increased tendency toward schizophrenia got an imaginary playmate? Good luck to you. Your manic-depressive-potential child has the giggles a lot. What does that mean? The kid with the tendency toward alcoholism—do you introduce him to the responsible social use of alcohol in the home, or throw out every bottle and declare your whole family teetotalers? Which way will that go? Rebellion? Does the kid have the genes that increase rebelliousness?

It gets ludicrous, fast. Except maybe when it's two o'clock in the morning and it's your kid.

Leah, my middle kid, had a birthmark on her belly. A raised, almost black spot. Comes with a somewhat increased risk of becoming malignant, I forget the numbers. Not a really high risk, but not trivial, either. Or not trivial to *me*. My husband and I decided to have it removed. We waited till she was seven, old enough so that it wouldn't be unduly traumatic. We could stay with her and talk her through a scary outpatient procedure, comfort her through a few days of soreness. But we did it early enough so that it was our decision, not hers. I'm pretty big on children's rights, and controlling their own bodies and all that. But I took care of this. I didn't want to spend the rest of my life worrying. I didn't want to be nagging her to have it checked. I didn't want to worry that she'd decide it was cute (it was) and display it in a bikini, evil rays poking at it right through the shredded ozone layer.

That was an easy one. What they're talking about now, with long lists of predictions and probabilities, and nothing much you can do about most of it, is terrifying. We talk about knowledge as being empowering. But this isn't knowledge. It's information, bits and pieces of possibilities. And it's not empowering; it's incapacitating.

It doesn't rest on how good or how bad the genetic science is. This isn't a technical problem. These moral dilemmas showed themselves early on. There was the big XYY fiasco. Someone discovered that men with a second Y chromosome, an extra one, seemed to be overrepresented in prisons and jumped to the conclusion that there was a "criminal gene" there. The data was lousy. The dilemma was real. Should parents of XYY boys be told? Do you have a right to keep this information from them? Do you have a right to give this information to them? What were they supposed to *do* about it? Some people who learned XYY from prenatal diagnosis (and you cannot look for an extra 21st chromosome without counting the Xs and Ys) terminated pregnancies. One man who got that diagnosis for his wife's pregnancy and chose with her to abort said, "It's hard enough to raise a normal kid. If he throws the blocks across the room will I think he's doing it because he's two or because he's XYY?"

The information, or pseudoinformation or whatever you want to call it, was incapacitating. He couldn't figure out how he was supposed to raise this child. I spoke to others in similar positions. Told predispositions, tendencies, possibilities in their potential children, they couldn't go through with the pregnancy. Not knowing what they knew. Or thought they knew. It was altogether too overwhelming a responsibility.

And what is this about planning the child's upbringing, organizing the diet, the exercise, the medical treatment? What comes to mind is something like a kid with a predisposition to diabetes, so you discourage sugar and candy. That might work. Might not, actually—who knows? Or a kid with a tendency toward lung disease: tell her not to smoke. That should take care of it. Surely a 12-year-old who knows she is at risk for lung disease wouldn't smoke.

Or maybe not, come to think of it. We've been telling all of our kids they're at risk for lung disease from smoking for quite some time now, and they're smoking away at higher rates than ever. Do we turn to two 12-year-old kids and say to one, "Don't smoke." And to the other, "*Really*, don't smoke." It sounds like the way I yell at my kids when I'm exasperated: "Stop that! I mean it!" Like maybe I didn't mean it the last six times I said it.

There is a limit to how much parents can control their children's diet and exercise and whatever. Plenty of parents would feel good if they could just keep their kids from guns, drugs, street violence. Plenty of people—even presidents—have found it hard to control their *own* diet and exercise. Not all people with adult-onset diabetes can control their diets. There are people with lung cancer or emphysema who cannot quit smoking. I'm not sure how much better we are going to do with individualized risk assessments passed out at birth.

And those are the risks that are, at least theoretically, under individual control. Most of our risks are not. You cannot construct a healthy environment one person at a time, child by child. People cannot simply decide to live in safe places, breathe good air, drink clean water and have it happen.

What can we find with this map? We will have positioned parents, mothers especially, between a world they cannot change and a life they might bring into it. What are their choices? I could remove the spot from Leah's belly and make the sun safer for her. If my child is at risk from the sun, air and water, if my child is at risk in ways I cannot control, then what? Where can this map take me?

The push has been to get this map of a child before the child. Prenatal diagnosis is pushed back earlier and earlier in pregnancy. Preimplantation diagnosis, doing the diagnosis on embryos outside of the body and only implanting the ones that pass muster, is yet another possibility. Increasingly we will know whatever there is to be known from genetic testing early, before birth, maybe even before the pregnancy. Maybe we can get this magic printout as we get a confirmation of a pregnancy, or maybe the printout will be there to guide us into a pregnancy, picking and choosing between embryonic possibilities. What will we know? The diseases, certainly. But also the predispositions, tendencies, increased or decreased probabilities toward all kinds of conditions, behaviors, types and events.

With later prenatal testing, the question has been how can a woman bear to end a wanted pregnancy. With earlier testing, we face a new question: how can you bear to continue? How much can you know, and still go on?

Beyond Prenatal Diagnosis

Prenatal diagnosis has been enormously costly. It has meant that women had to begin pregnancies, go through weeks or even months of being pregnant, caring for themselves and their babies, all the while being prepared to end the pregnancy. They had to have tests that may themselves have put the pregnancy at risk. And then, if there was bad news, they had to go through late abortions.

When I first began studying this, amniocentesis was the technology in use. The woman had to lie still while someone poked a long needle through her pregnant belly, into the womb itself, and withdrew some fluid. That fluid was then cultured, and only then could the fetal cells be diagnosed. The fluid couldn't be withdrawn until at least 18 weeks into the pregnancy and it took several weeks to culture. Results typically came back after 21, 22, 23 weeks. A full-term pregnancy is 40 weeks. Abortion was legal for 26. By the time a decision was made, the woman looked pregnant, felt pregnant. Not only could the woman herself feel fetal movement, but by that point you could see it right through maternity clothes. It was hell.

I made myself very unpopular talking about it. Women went through this believing they were sparing their babies from suffering. And who am I to say what a given disability would have meant to that baby, to that mother? So some disability-rights activists were angry at me. Women going through it talked about their babies, not their fetuses, or "products of conception." They were acting as mothers, taking grief and responsibility upon themselves in what they felt was the interests of their children. So they used the language of infanticide, not abortion, and some abortion-rights activists were angry at me. But angriest of them all were the doctors and the genetic counselors. They were giving women information and sparing them tragedies: and what *was* my problem?

The women were my problem. Their voices haunted me. They were grieving mothers, and people kept telling them how lucky they were. People talked about the tragedy these women had avoided; the women told me about the tragedy that had happened.

I was invited to speak at a continuing education program for physicians in Aspen. Seminars were early morning and late evening, *après ski*. It was after nine in the evening when I faced a room full of physicians, their faces rosy from their day's skiing, sprawled in comfortable armchairs, drinks in hand from the open bar in the back of the room. I thought I'd tell them what the women were going through. I'd share the stories of how doctors, along with husbands, friends, colleagues and passing acquaintances, made this worse. Maybe, if they heard this, the doctors would handle it better.

I told about the woman whose doctor got the bad news the day before her regular checkup. He had her come in, lie on the examining table, set up the doptone to listen to the baby's heartbeat, and then asked if she'd heard yet from the genetic counselor.

> Meanwhile the baby's heartbeat is pounding, you hear it over the speakers, and I said, "Why, did you hear from her?" He said, "Yes, she tried to get you today." Now I figured. I said, "Is something wrong?" He said, "Yes." All the time the heart is beating overhead. I said, "What is it?" He said, "I'll tell you in a minute." I said, "Tell me." He said, "Well, we call it Down syndrome." And the heart is beating away nice and healthy.

Her world stopped, crashed around her ears, and he went on checking the health of the baby she was about to lose.

I told them about women who were informed over the phone and told they needed to decide and to schedule the termination in that same phone conversation. I told them about women whose doctors wouldn't tell them what to expect, didn't know what a late-term abortion would be like, and were angry about that. And I told them about women who were *very* well informed, bludgeoned with information. One woman's doctor explained to her that the abortion "is almost on a hairline when babies start to survive," and then told her "because my uterus felt big to him, he said that they wanted to do the saline just to make sure that the baby wouldn't be born alive." And then later, she told me, "The point of most pain for me was when the doctor injected the saline and I knew that that was the point when I was killing the baby. That was, for me, the most difficult point."

I told them all of this and more. About the grieving, the loss, the months of crying. I told them about women who couldn't go through this again, and couldn't *not* go through this again, so chose never to get pregnant again. "No more babies."

I talked and they listened. I'll grant them that. The room got real quiet. They sat up and listened to me. And when I was done, they let me have it. They didn't know who I'd been talking to, or where I'd found these women, but I didn't know what I was talking about. If I want to see tragedy, I should see the women who didn't have amnio and had babies that were real tragedies. Then I'd know. One doctor was really angry at me. "My women," he shouted, "my women are grateful! They don't cry over this. They are grateful!"

When it was over, I made my way to that bar in the back of the room. A woman approached me. She was the wife of one of those doctors. "He didn't understand when I had my miscarriage. How's he going to understand this?"

That was the end of my career in continuing ed for docs for a while. But then, a few years went by, and new technologies came in. While lots of women still go through later amnios (a typical scenario is having a blood test at about 16 weeks to determine the risk level, repeating that perhaps, and then going on to an amnio as late as the ones had been years before), it is increasingly possible to do the amnio or other diagnostic testing earlier. Early amnio is done at 10 or 12 weeks, chorionic villus sampling even a bit earlier. With the new technology in hand, it suddenly was okay to talk about how bad the old amnios were. I found myself invited to conferences again, this time placed right before someone who described the newer, earlier tests. Another problem solved.

And it should be, it just *has* to be better to do this earlier. But better isn't good, and grief doesn't lend itself to accounting procedures.

I used to get phone calls from women who were going through this. They'd read *The Tentative Pregnancy* and tracked me down. They'd call and tell me their story, wanting to have somebody listen who understood. So I listened. One day a woman called me at home in the evening, and the story was familiar. I listened, said soothing things, did what I could. She herself was a physician, had the test, got the news, went right back to work a few days after the abortion. But then she couldn't work any more, couldn't stop crying. Her husband kept telling her to snap out of it but she couldn't. She was so sad. Her baby—she kept thinking about how old the baby would have been, when the birth would have happened, what the baby would be like now. I listened, I murmured appropriately. As we were saying our goodbyes, she said, "Just think how much worse this would have been with an amnio." She'd had the earlier tests, the earlier abortion.

Prenatal diagnosis is costly, and it is crude. You have to get pregnant, check the fetus, and then either keep it or not. There's not much choice to be made, caught betwen bad and worse, whichever is which.

If this is just too hard, if you cannot keep asking women to go through this, then what are our options?

One is the almost unthinkable: go back. Go back to where we were, stop doing prenatal diagnosis, and deal with the babies as they are born. For most of the conditions for which prenatal diagnosis and selective abortion are used, we would have to make a collective commitment to making the world better for people with disabilities. We would have to take collective responsibility for the added costs— not just, but certainly at least, the financial costs. That most assuredly is not the direction in which our society has been headed, but it is the direction we *ought* to be headed, with or without prenatal testing.

But even a utopian social world would still leave us with a few babies born whose condition is best described as suffering. There are babies who are experiencing great pain, babies who are dying badly, babies for whom life itself seems a great and cosmic mistake. Some of those babies are being born, are suffering, are dying right now, in spite of the existence of prenatal testing. No matter how much testing we do, some such babies will be born. Stop testing, and some more such babies will join us. What do we owe them? How can we care for them? Are we ready to reopen the discussion of euthanasia, of "mercy killing" not with a semihidden eugenics agenda, getting rid of "defectives," but as an alternative to the more massive eugenics project that prenatal diagnosis threatens to become?

I first started saying this in graduate seminar rooms, behind closed doors. It shocks, but maybe less than I thought it would. People who've looked at the way people die in American hospitals and begun to think the Kevorkian approach has some merit as compared to intensive care units, may be ready to think about infanticide the same way—not as the eugenicist's "black stork," killing the defectives to save the race, but as a welcome and timely visit from the angel of death, a release. But maybe you *are* shocked.

So what is the other alternative? The yet-higher-tech alternative that opens up is not to go back, but to go ahead, to use yet more sophisticated technology. Rather than diagnosing fetuses or even implanted embryos, the other option is to get in earlier. With the new technologies of procreation, the technologies we have heard so much about in the treatment of infertility, embryos can be made or can be moved outside of the body, and diagnosis can be done on just a few cells, before implantation. Or, even earlier on, we have the futuristic model of genetic engineering: don't wait to develop and then discard unacceptable embryos, but rather make only the embryos you want.

So the brave new world opens up. There are thousands of perfectly good, healthy people in the world with wonderful characteristics. Why not just clone them? Or why not pick and choose among characteristics and put together the baby-of-choice from the assembly of discrete dominant and recessive alleles available, no longer randomly assorted?

Then the selection begins to switch over, from selecting against the specific condition we, rightly or wrongly, fear, to selecting for characteristics we, again rightly or wrongly, want to have. If you are choosing your embryo out of a catalog of available cloned models (what George Annas has presented as the "Cabbage Patch" model) or in a slightly less high-tech fantasy, choosing among the dozen embryos produced in your own in vitro procedure, of course you choose against the characteristics you don't want. But what about choosing for the characteristics you do want? Some couple has two healthy boys and because of the mother's age is now at risk for Down syndrome, or learns of some other genetic risk. In this third pregnancy, they choose to use one of the in vitro procedures for preimplantation diagnosis. The embryos are checked and those that carry the conditions they seek to avoid are discarded. Eight acceptable embryos are left, and you cannot implant them all. Should they choose at random? Or should they choose the female embryos, have the daughter they've always wanted?

What happens when we remove abortion from selective abortion, and face only the selection process?

Now You Can Choose!

If conditions incompatible with life are the moral high ground for
bioethicists confronting the questions raised by prenatal diagnosis, where is the point where things get iffy? Any study of the introduction of new technology has to confront the problem of the
technological imperative. Once people know how to do something,
it is very hard not to do it. And once people know how to do something, they do it more and more, often with wider and wider applica-
tions. If here, why not there? If then, why not now?

In philosophical terms, this problem is often discussed as the
"slippery slope," a long slide down from acceptable to (morally) unacceptable uses of technology. In the case of prenatal diagnosis, the
slope is usually graphed as moving from diagnosis and abortion for
conditions incompatible with life, passing through the firm but con-
tested territory of Down syndrome and neural tube defects, floun-
dering on the rocky terrain of socially undesirable conditions like
deafness on down to obesity, bouncing along the questionable areas
of "manic-depressive genes" and "alcoholism genes," and finally
crashing into the great moral abyss of sex selection.

Now *that*, the bioethicists generally agree, is wrong. Using prenatal diagnosis and selective abortion "just because" you want one sex
and not the other is generally considered wrong—not only by
bioethicists but by lots of ordinary folk too. It has become, I think, a
fairly standard place to "draw the line." For people who are doing
something that risks making them uncomfortable, morally edgy,
having a line somewhere is reassuring: this here may be a bit tricky,
but that there is *wrong*.

I've heard countless physicians, geneticists and genetic counselors use sex selection as the line, the unacceptable place to go,
which by its very existence makes what they are doing more acceptable. I am very uncomfortable with that line. It bothers me because it
seems to make two demarcations, neither of which feel right to me.
One is the line between "medical" and "nonmedical" conditions.
The argument is that prenatal diagnosis and selective abortion for a
"medical condition" is morally acceptable. The original language,
which one still sometimes hears, is that those are "therapeutic" abortions. The implication is that a medical decision is a scientific, ratio-
nal decision, and one that is morally sound because it is in the
interests of health.

But what exactly makes a given disability a "medical condition"
and sex *not?* Sex is a diagnosable genetic condition, associated with
variations in phenotype, health, longevity, life chances. Is it that sex,
while it makes a difference physically, should not matter that much
socially? Is it that both sexes should have what they need to have a
good life? Is it that sex ought not to be a basis for valuing people?

Then what is the difference between sex and another "genetic"
condition? This is precisely what disability activists are saying: deaf
people, people in wheelchairs, people who are blind, people with retarded mental development, all ought to be given what they need to
have a good life, and all ought to be valued for what and who they
are. The problem, the disability activists are saying, lies with the society. And so it is, one can easily argue, with sex. We ought to value our
girls and our boys, welcome both equally. There is something wrong
with a society that feels the need to do sex selection.

And that brings me to the second demarcation that makes me
so uncomfortable: "us" and "them." Over and over again, I've heard
American and European physicians and geneticists point out that
sex selection is a "Third World" problem, something done "over
there" and requested by "immigrants." *We*, the doctors tell conference audiences, use prenatal diagnosis and selective abortion for
sound medical reasons. *They* misuse the technology.

Each and every woman who uses this extraordinarily difficult technology of selective abortion is making a decision based on what she knows about the baby-to-be, and what she knows about the world into which she might bring that baby. An Indian woman who knows what faces her third daughter is not making a morally different decision, it seems to me, than an American woman who knows what faces her child with Down syndrome. I've talked to women in both of these positions, who spoke with great love and longing for the baby-that-might-have-been, and much regret about the world that is. Women in each of these positions have said to me, "It wouldn't be fair to the baby, or to my other children."

I'm not trying to show you how similar selective abortion for sex or for disability are in order to show you that sex selection is right or that disability selection is wrong. Rather, I remain the sociologist: we need to look not at the individual decision, but at the social context, the world in which that decision is made. I can understand and respect a woman who chooses to terminate a pregnancy based on what she knows about the fetus and the world. I cannot understand or respect a society that puts her into that position when it is not inevitable.

Let's take it beyond sex. Consider race. Race too is a "genetic" condition expressing itself in phenotype, health, longevity and life chances. Whatever characteristics of skin, hair and bone that a particular society defines as "racial," those characteristics will write themselves upon the body of the child and be used to shape its life. Years back, feeling somewhat defensive on issues of disability, I was trying to explain the position I was taking. It was not that the women were themselves "antidisability," "ablist," not that they themselves necessarily felt any repugnance toward people with disabilities when they made the decision to terminate a pregnancy with a fetus so diagnosed. There was one woman I interviewed who expressed the dilemma forcefully. She was a teacher of mentally retarded people, and had been for 13 years. It was her life's work. She said that she was tempted to go ahead and have her diagnosed Down syndrome baby, and show the world what a wonderful job she could do raising it. But she knew all too well what happens to these children in America when they grow up, when their parents die, and she wasn't going to put a child of hers in that situation to prove a point.

Groping for a comparison at a college lecture, I posed this scenario: Consider a South African white woman still living under apartheid. Such a woman, pregnant by her black lover, might choose an abortion rather than bring a black child into that situation. Does that make her a racist? I thought not: I thought she was doing what she felt she had to do as an individual living under impossible circumstances. If we were going to have discussions about regulating morality, it wasn't *her* morality I'd start with.

A young black man came up to me afterwards. But what about *black* South African women, he asked. What were they supposed to do? They too are bearing black babies into that world. And indeed, what of those women? Haven't they sometimes made that same decision? It's not always about the *otherness* of the baby. Black women, Jewish women, women under a variety of racist regimes have contracepted, have aborted and have committed infanticide to save their babies the consequences of racist madness. When a woman kills her baby rather than having it sold downriver to speculators, or starved, tortured or experimented upon, isn't that a mercy killing?

And what of women with disabilities? They've not been spared these decisions. Some years back, in one of the excesses of "talk radio," a show focused on the situation of a television anchor's pregnancy. The woman had a genetic condition which resulted in missing fingers and toes. Her hands were affected. She was pregnant, and had the possibility of passing this condition on to her child. Some callers felt she had no right to continue the pregnancy without testing. The woman herself obviously wasn't all that severely disabled—she was, after all, a television anchor and so by whatever American standards one brings to bear, doing just fine thank you. But other women have been yet more disabled, have found the world too hostile a place for people with their condition, or their condition too difficult in itself, and have indeed chosen not to bring a child like themselves into the world.

When selective abortion was the only option, there was always and inevitably a strong leaning toward continuing the pregnancy. You have to have a good reason to abort a wanted pregnancy, and to some extent, the later the testing, the better the reason had to be. So the moral questions that arose were about what warranted an abortion: which conditions were serious enough, which too frivolous? It turns out not to be so straightforward after all to draw a line that neatly divides "medical" and "other" grounds, that separates us with our good decisions from them with their bad ones. All decisions are made in a context, and there is no objective place to stand and judge.

But the technology is not stable. It shifts, and with the shifts come different questions. When you are not selecting *against*, but selecting *for*, the issue changes.

Selective abortion has been a "slippery slope" problem. But when we think about selective implantation, selective *creation*, then I prefer a different image, that of the "camel's nose." That argument goes that once you let the camel's nose into the tent, it is very hard to keep the rest of the camel out. Instead of a slippery terrain along which we societal explorers move at our own risk, I see a very aggressive camel: the biotech industries, highly motivated to get the nose and the rest of that profitable camel entrenched in our tents.

Sex selection is a perfect example, with clinics opening up around the world. There are techniques of selection, of choosing the sex of a baby, that do not involve abortion. One is sperm sorting, using the technology of "artificial insemination" after separating out X- and Y-bearing sperm. It's of dubious merit as a technology, only promising increasing odds from roughly 50-50 to roughly 70-30. But thousands of people have paid for it, have tried it. There are other, more elaborate techniques, with higher success rates, combining variations on in vitro fertilization. Embryos, outside of the body, can be sexed and only the "right sex" implanted.

These are all technologies that are being marketed, that are there for a price. It is also possible to choose the *race* of your baby. Great Britain had a notorious case of a black woman requesting an egg from a white woman for in vitro fertilization, so she could spare her child the burdens of racism.

Sometimes people do sex selection to avoid the consequences of deep and profound sexism in their societies, and then it is, I think, comparable in every way to decisions women make about terminating pregnancies for disabilities.

But, especially when abortion is not involved, sex selection is also offered as a "consumer choice." Depending on how invasive a procedure she is willing to have, and how much money she wants to spend, a woman can choose a method and at minimum increase her odds, and at maximum virtually guarantee having a child of the chosen sex.

We are introducing choice into yet another arena of our lives. Choice always seems like a good thing to have. And from the point of view of the consumer, the purchaser, it probably is. But what about choice from the point of view of the—what? consumed? purchased? Let's be kind: the *chosen* child.

People in the adoption world have struggled with the idea of the chosen child. At first, it seemed such a satisfying bedtime tale: we chose you. It seemed, in adoption, a nice counterbalance to the implicit, understood but unstated fact of having been un-chosen, having been placed, made available in the first place.

But it wasn't such a sweet tale after all, and adoption workers now discourage using that story. If a child is chosen, it is chosen *for* something. Why me, the child asks, what about me made you choose me? And suddenly parenthood becomes contingent. Chosen for being pretty, sweet, cute, for any given characteristic, means if you lose that characteristic, your chosen status is at risk. What if I get ugly, surly, stop being cute? the child has to wonder.

But if you choose a boy and get a boy, or choose a girl and get a girl, then what's the problem? The technology can be improved so that failure, the wrong sex, is not going to happen. So then won't the child have been chosen for what it truly, indelibly is?

Who uses sex selection? A woman who has two sons and has always wanted a daughter. A family that has all girls and wants to "pass on the family name." People want sex selection because they want a particular kind of child—or maybe, more accurately, because they want a particular kind of parenting experience.

Sex is a diagnosable chromosomal condition. Choose for Y and you get a male; choose for X and you get a female. But what is it, exactly, that you are choosing and getting?

Sometimes people call it "gender selection," which is technically wrong, but in a larger sense quite right. "Sex" is the word social scientists use for the biological phenomenon of male/female. "Gender" we save for the social role, for being a boy, a girl, a man, a woman. A sperm cell or a zygote can't possibly have "gender," but gender is what people are choosing when they select by sex. People who say they want a girl have something in mind: girlness, femininity, some set of characteristics that they expect will come in that girl-package that they think wouldn't come in a boy. I've heard women say that they want the kind of relationship they had with their mothers; they think they can't have that kind of relationship with a son. I've heard women talk about wanting to have the frills, the clothes, the manicures together, the pretty mother-daughter outfits, the fun of a prom gown and a wedding gown, that come with girls.

Can't you just see the disaster looming? That woman is not ready for a six-foot-tall, 300-pound daughter who wears nothing but denim and boots. People who want a son are probably none too pleased when he announces he wants ballet lessons. When people want a son or want a daughter, they want a host of characteristics that they believe are, and often believe *should be*, sex-linked. Someone who wants a cuddly, warm, loving child who will remain close chooses a girl. Someone who wants a child who will go out and accomplish great things in the world, make a name for themselves with the family name, goes for a boy. The person is choosing sex, the chromosomes and the genitals, but they are also making a statement about personality, lifestyle, what they want their children to be and to become. They're opting for gender.

Sex is a very crude determinate of these personality and lifestyle traits. How much of gender is biological, "genetic," "nature" is a long-standing debate. But wherever you stand on that question, it is apparent that not every child is a perfect gender stereotype. Getting a girl or a boy doesn't guarantee a parent the characteristics he or she is seeking.

Well, lots of us aren't what our parents had in mind, and so be it. But this technology offers an implicit guarantee, a promise that parents can choose, can hope to control the kind of child they will be parenting.

Determination

What are the characteristics that we think we can control when we plan our children? When we've moved past the list of diseases to be avoided, then where are we? Sex is a crude genetic characteristic: it is writ as large as a chromosome, and is only a loose indicator of what we might expect in a child. As the map is unfolded and read, we expect finer and finer resolution. But we're looking at a map, not a crystal ball, and we're not dealing with three wishes from the blessing fairy. Our planning is limited to selecting among embryos, or selecting specific stretches of DNA to include in an embryo, and so our choices are limited to those things we believe are genetically determined.

The logic of genetic thinking as Evelyn Fox Keller summed it up is that genes are primary agents of life; they are the fundamental units of biological traits. According to that logic, to read genes is to predict traits; and to order traits you have to select or construct the genes. Want a blue-eyed kid? Select for the gene that causes blue eyes.

But are genes causes? Ruth Hubbard would have me never use the word "cause" for the action of a gene. She went through an early draft of this book with a red pen, and took out every "cause" phrase, every bit of "gene for" language. A gene does not cause or do something, and it certainly is not "for" something. A gene is associated with, involved in, active in. And while I know she is right, somehow I think that the language of a gene that causes something, like blue eyes or sickle-cell anemia, is a reasonable way to speak. What, after all, ever causes anything in this world?

I fell down the stairs and broke my ankle. But it is perhaps a bit more complicated. How did I come to be on the stairs? My aunt Joan lent us the down payment for this house with its big staircase. I was carrying the Chanukah presents at the time and couldn't see where I was going—Judah Maccabee fought the battle that Chanukah commemorates. American Jews only make such a fuss over Chanukah to compete with Christmas. Actually, to compete with the commercialization of Christmas. And besides, some little child who shall remain nameless lest godforbid she get a complex, left a plastic bag on the third-from-bottom step. And how can you break both bones in your ankle by falling three steps? Look at a skeleton some time—the whole weight of the body tapers down to this absurdly thin point right above where the foot twists.

So what was the cause of my broken ankle? Aunt Joan, Judah Maccabee, Jesus Christ, American capitalism, a nameless child and an orthopedic design flaw.

"Causality" in science is basically only a hypothesis you can't disprove (yet). So with the hedging of my bets and all, I'm still ready to use the word "cause" sometimes in connection with a gene. I know that genes only code for the production of proteins. They don't *do* anything, they don't even produce the protein, but still, in the more-or-less approximate way I use to talk about cause, I feel comfortable saying that a gene causes, say, blue eyes. Sickle-cell anemia. What else?

For the longest time, I resisted the suggestions that I write about genetics because I wanted to avoid precisely this discussion. We've walked ourselves into a corner with this nature/nurture thing: we've set up a dichotomy that exists nowhere but in our own heads, and then keep confronting it as if it were a fundamental truth of the universe. We make a list of characteristics, qualities, traits, states of being, and then see if we can assign them to the "nature" side by finding a "gene for" the characteristic. Is intelligence, sexual orientation, schizophrenia, the tendency to divorce, depression, the inability to spell, genetic or not? The discussion too often seems to degenerate into "Is too!" and "Is not!"

Probably the most public, vociferous, and politically important of the discussions has been the long-standing one focused on genetic components in intelligence, and the more recent one focused on the "gay gene." The intelligence discussion has been hopelessly mired in the racism that surrounds it. So let's take a look at the "gay gene" discussion: Is there a "gene for" being gay? For gay men, there does seem to be a genetic component, one of the pieces. Like every other gene, it speaks in probabilities, in odds. If an identical twin is gay, the chances of his twin being gay are 50 percent. That is considerably higher than chance, but an awful lot lower than the odds on eye color.

Chandler Burr has helpfully compared male sexual orientation with handedness. For both we have a dominant and a minority orientation. About 92 percent of the population is right-handed. Left-handedness has at various times in history been treated as evil, sick, sinful, or an ordinary variation. Handedness is experienced as a very powerful given: it is not changeable by an act of will, though one can hide or pass if necessary. And so it seems to be with male homosexuality.

But is handedness "genetic"? If an identical twin is left-handed, the chances of his twin being left-handed are 12 percent, or one and a half times chance. That is, the identical twin of a left-handed person is only one and a half times more likely to be left-handed than is the person sitting next to him on the bus. It is not a powerful argument for genetic causality. But it doesn't make handedness a "lifestyle choice."

It seems as if what we are really talking about when we invoke genes is predestination versus free will. We often talk as if the opposite of "genetic" were "a choice." Genes seem to function in our language and our thinking as equivalent to inevitability, determination, predestination, fate. Yet what we hear from the geneticists is that genes work as probability factors in a causal equation. They play their part.

If you keep leaving things on the steps (and if I've said this once I've said it a thousand times), someone's going to get hurt. And if people march up and down stairs day after day, year after year, it's no surprise when eventually someone falls. And if you carry packages and can't see where you are going, well, what do you expect? To each of these things, and probably a dozen more, maybe even one or two that are "genetic," having to do with bone structure, clumsiness and distractibility, we can assign a probability rating. What are the odds of falling under each set of circumstances?

Now we're approaching some basic philosophical questions about determinism and inevitability. Was it inevitable that I break my leg? Given everything that happened in the world to that exact second—including the history of architecture, my relationship with Aunt Joan, the world history that brought that nameless little child who shouldn't have a complex about this into my life, the invention of plastic, the evolution of the ankle—given all of it, was it inevitable? Did I have to put my foot there? Was it fate?

That is a fascinating philosophical question, but it is not terribly useful practically. For practical purposes, we focus on one or two of the factors that we think we can control. Don't leave things on the stairs. Watch where you are going. And we act as if—we have to act as if—we have control.

When genes become more and more important in our thinking, we start assigning them greater and greater causal power, moving them to more central positions. Sometimes that has meant giving up, that metaphorical throwing your hands up in the air and saying "it's genetic," meaning "And that's that." Which is okay, if the situation is one that we might want people to take their hands off of and leave be. So the "gay gene" might be useful as a political tool if invoking that gene becomes another way of saying give it up, you have to accept that some people are inevitably, determinedly, gay. But if the question we are looking at is not "Why are some men gay?" but "Why are more black men in prisons than in colleges?" then saying "It's genetic" is quite dangerous. See *The Bell Curve.*

What complicates it these days is that the way things are going, "It's genetic" might very quickly not be a throwing-up-of-your-hands kind of problem, but a rolling-up-of-your-sleeves kind of problem. "It's genetic" might be coming to mean, so let's fix it, let's engineer it, let's construct it to order. Let *us* make the determination, and let us predetermine.

Take that highly publicized "gay gene," XQ28, now officially recorded as GAY 1. Individual prospective parents of privilege should soon be able to include that—or any other given gene—in their list of things to select for or select against. Lots of people would be considerably more distressed to learn that their child is gay than to learn that he has some disability. The idea that there is a genetic component to being gay leads pretty quickly to either selecting against that gene or engineering to change it. I picture that aggressive camel at the side of our tent, wearing its sign advertising sex selection: won't heterosexual orientation, for a slight additional fee, be available in the newer package?

Gay is a highly politicized trait. But every day seems to bring some other "gene for" some other quality, characteristic, trait. Can we control all of it? Can we test and select, read and decode and splice our way to what we really want in our children, for our children?

And have we any right to do that? I'm not talking about our legal rights, our rights as citizens. American liberal legal scholars can show that whatever our discomforts with treating children like consumer products, it is not in our civil libertarian tradition, and probably not in our interests, to try to stop each other from doing so. So I'm not going to argue against "choice," not even the very limited kind of choices that are available to some people and not to others in so profoundly unjust a society. Rather, I am thinking about our rights as parents in our relationships with our children. Do we even want to order them, to have them custom-made? Would we have wanted our parents to have ordered us, chosen our traits, predetermined whatever they could or wanted to about us? Whether it is what you like best or least about yourself, you probably won't like thinking about that as something your parents put on an order form.

Parenthood does not come with guarantees. Motherhood, I've often said, is one more chance for a speeding truck to ruin your life.

The world has plans for your children, and your children have plans for themselves: you will not be able to control this.

The demands of the information age drive us toward getting all the information, toward taking all the control that we can. Perhaps wisdom lies in not always doing so, in making wise judgments about what information we want, and what information we do not want; which choices we want to make, and which choices are not ours to make.

Motherhood is not about consumer choice, and it's not about guarantees. The Dutch word for midwife, *vroedvrouw*, translates—as it does in many languages—to "wise woman." For wisdom on motherhood, I turn to midwives. The Dutch midwives worry about all of the new testing, and the implicit but false guarantee it seems to offer. Women say they want testing because they could "never raise a retarded child," or "couldn't bear" to have a disabled child. As if these things could be predetermined, as if they were all written in code ahead of time. Test, select, do what you think you can, but remember, as one midwife said,

> You are eager to have a healthy child, but after a chorionic villus sampling, amnio, an ultrasound and birth, your worries are not over yet. When the child is there you still have your concerns. Can he walk along the street on his own, and near the water, I hope he gets no accident, and I hope he doesn't get some wrong friends. It is a process, all life long, isn't it? Somehow or somewhere you have to let it go, you cannot control everything, and maybe you have to start to let it go at the beginning. You should dare to leave some questions without an answer.

On Order

The technologies of procreation are about control: from the simplest of contraceptives, through in vitro fertilization, amniocentesis, embryo biopsy and preimplantation diagnosis, all the way to the newest twist, cloning. Cloning is quite markedly about control. It's about introducing predictability and order into the wildly unpredictable crapshoot that is life. If normal procreation is the roll of a hundred thousand dice, a random dip in the gene pool, cloning is a carefully placed order. And that's where it gets interesting: it is *order* both in the sense of predictability and control; and it is *order* in the sense of the market, a (human) being on order.

In a perfect world, we could think about the value of the first form of order, the value of predictability and control in procreation, without thinking about the second form of order, the power of the market. But in our world, the two are hopelessly, endlessly entangled.

We've already moved in the direction of the marketing of babies, even without the technological sophistication of cloning. There is of course the adoption market, which often compares unfavorably in terms of decency and honesty in human relations with the marketing of used cars. But marketing is also operative in the technologies of procreation. If you want to purchase sperm or eggs, you can get listings of the available merchandise. The characteristics of the donors/sellers/producers of the genetic material are provided. Some of the things listed are fairly direct predictors of likely genetic attributes: height, hair color. Some are in the far more questionable areas: intelligence, personality traits, talents. Some are downright strange, like religion and hobbies. Think there are genes for Methodists or stamp collectors?

But sperm or eggs are a long way from people. It's even a long way from the genetic attributes of the people that might grow from those gametes. Someone who has any given gene might not pass that gene on in the particular gamete the purchaser ends up with. People who use sperm from the "genius sperm bank" have to understand that their kid might inherit the genius's nose, not his brain. And a gene that is passed on may or may not express itself when in combination with the other gamete with which it pairs.

Cloning moves control up one more level. There are two things that are sometimes called cloning. One is what the scientists mean by true cloning: an adult cell is used as the nucleus of an egg. The being grown from that egg is a genetic twin of the adult being from whom the cell was taken. That's the form of cloning that made the news with Dolly, the sheep made by cloning in Scotland. That's the form of cloning people have in mind when they talk about creating another Einstein or another Hitler out of their very cells.

Sometimes the word "cloning" is used when what is meant is really "twinning," in which embryos are split and can be artificially split over and over again, producing vast arrays of genetic twins. That's what they do with cows all of the time: embryos produced by expensive cows are twinned in quantity, and then placed in the uteri of cheaper, more disposable cows, to be bred. Given enough time, there's not much difference between the two techniques. If a sample of the embryos is grown to adulthood while the rest are frozen, you can still end up with genetic twins of vastly different ages. From a marketing point of view, that means you can show people the grown merchandise before they purchase the kit. When selling to dairy farmers, you can show them grown cows of known levels of milk production, and sell the embryos that are their twins. I think of Julia Child as she's appeared on television, putting the bread in one oven and walking across the kitchen to show us the finished product coming out of the other.

We already know that the two loaves, even in Julia Child's capa-

ble hands, will never be identical. Predictability and control are forever slipping out of our hands. When you're working with small
herds of sheep, cloned to produce some expensive exotic protein in
their milk for medicinal purposes, as Dolly was, a certain percentage
of error is to be expected, accounted for. It is accounted for in two
senses: it is part of the expectations, the "account" or narrative of
what happens. And it is accounted for in the ledger books, an anticipated expense. Most of the errors that can be expected will not matter: only a few will affect the single purpose for which the sheep were
cloned: the production of that protein in their milk.

With people, the accounting is a lot more complicated, in both
senses. Errors are not to be written off, and our expectations are
rarely so narrowly confined.

Genuine human cloning, cloning from adult cells, is probably
not the direction in which we are likely to go. People do find it dis-
tasteful. It might well serve as a convenient boundary line, defining
what we don't do, which will make what we do more palatable.

With twinning, producing sets of identical embryos, one could
choose an embryo from a variety of models. With the addition of the
genetic technologies that permit DNA splicing, specific characteristics could be included, permitting and maybe even encouraging
some customizing. (I'll take that one, but with red hair and no dimples.)

Could such a thing ever happen? We know, those of us who have
been around the track a few times, how this latest "advance" in
human procreative technology is going to be brought to us: it is
going to be the solution to some heartrending problem. There's
going to be a very good reason to do it the first time: some baby conceived through in vitro fertilization, with an embryo twin still in the
freezer, is going to need a bone marrow transplant. Or maybe it will
have to do with "isolates of historic interest," communities of people
dying out. Or maybe.... But I don't want to give anybody any ideas.
Do I think we could end up with catalogues of people the way dairy
farmers have catalogues of cows? Why yes, I do.

How could we get from a set of technologies of procreation that were designed to end pregnancies where the fetus would suffer and die, to a set of technologies that turn babies into consumer objects? Was it the slippery slope or the camel's nose? Or was it, as my mother taught me, that one thing leads you into another?

Once when I was a kid, I was supposed to do a quick straighten, dust and vacuum before my mother came home from work and company came to dinner. Just get the shoes, socks and toys out of the living room, dishes out of the sink, make the house "presentable." My little brother's muddy fingerprints were on a closet door in a tiny hallway. I took a cleaning rag and some Spic & Span and gave it a wipe. Wow. A big white spot. I got the mud off, but also the year's accumulation of dust, cooking grease, whatever else yellows walls between paint jobs. I cleaned off that whole door. And the frame. The other doors in the hallway looked awful. I did them. The rag brushed the ceiling over the door frames. My mother came home. The house was a mess, and what was I doing on a ladder cleaning the ceiling of the little hall?

Well, one thing leads you into another.

This way of thinking reminds us that people aren't quite so passive. Yes, it is a slippery slope indeed, and it surely is an aggressive camel. But I'm the damn fool with a rag on top of the ladder.

Still, I'm too much the sociologist, and far too affected by the women I've talked to who have used technologies of procreation to frame these issues entirely in terms of individual choice. Women making decisions about prenatal testing often use the phrase "my only choice," to mean the choice that they felt forced into, a no-choice choice. As individuals we are often without real choice. Even the scientists themselves who develop and introduce the various technologies are helpless to control it. Someone artificially twins cows or clones sheep and says, "Oh, we never meant for this to be used on people." But slopes and camels being what they are, one thing does lead you to another.

How do you come to find yourself on top of a ladder cleaning a ceiling? You let yourself move from one thing to the next without remembering what the larger project is.

In the area of reproductive technologies, some of us are hampered by not being entirely sure what the larger project is. It has to do with the meaning of life. And while I don't know what that might be, I'm sure it won't be served by turning the creation of people over to the forces of the market.

We began by selecting against very specific characteristics, dreadful diseases, horrific possibilities. And, some would argue, some not-so-dreadful conditions, too. But in so doing, we started splintering the self, started crystallizing the potential person into component parts. Once we've done that, removing the parts we don't want may not be enough. We might begin to sort through for the characteristics we do want.

In earlier times, the distinction was made between negative eugenics, the sterilization or even murder of "genetic weaklings," and positive eugenics, encouraging the "genetically fit" to breed. Once we start reading the code, we come to see that we are all made up of some genes we do want to perpetuate and others we do not. This microeugenics is no longer about groups of people: it is now about specific stretches of code.

I've said that we shouldn't think of these stretches of code as "blueprints" for the construction of a person. I've said to think of them rather as bread recipes, with all of the individual diversity and variation that growth and time introduce. But what happens when we turn such a process over to the forces of the market? Wonder Bread: a nearly perfectly predictable bread.

I cannot afford to do what my great-great-grandmother did out of necessity, bake my own bread, any more than I could afford her uncontrolled fertility. I can't afford the time. Home-baked bread, like eight children, are luxuries well beyond the likes of me. I almost always buy my bread, baking only for a treat, a holiday, for the occasional pleasure of it. I can afford to buy a more customized bread than Wonder Bread: I buy it at the bakery, at the farmers' market. I buy the services of the baker, considerably more industrialized than what I could do at home, somewhat less industrialized than the factory supplying my supermarket. And sometimes, when

I'm rushed or broke, I grab those plastic-wrapped loaves off the shelf.

Could people ever be mass-produced like supermarket loaves? I don't actually think so, but the image frightens. The factories toss out all the errors—the loaves that come out misshapen, that failed to rise evenly, that burnt a bit or were underdone. It happens. They take that into account. It is part of their quality control program.

The current technologies of procreation introduced first quantity control and more recently, and increasingly so, quality control. Prenatal screening and testing with selective abortion are a form of quality control, avoiding the production of the children we claim we can no longer afford to raise. The choice of quantity control, the choices offered to us by (relatively) safe and (relatively) effective contraception, eventually lost us the choice of not controlling the quantity of our children: who now feels they could really afford the eight children my great-grandmother bore? And so it is with quality control: the introduction of that choice may ultimately cost us the choice not to control the quality, the choice of taking our chances in life's great, glorious and terrifying roll of the 100,000 dice.

One thing does lead you into another. Cloning, or embryo twinning, splicing specific characteristics in and out, may eventually be offered to us as a way of avoiding the tragedy of prenatal diagnosis and selective abortion. Cloning may eventually—and eventually isn't as long a time as it used to be—be offered to us as a way of inserting predictability and control earlier in the process. Placing order in procreation: placing our orders.

In the hands of the market, the "book of life" becomes a catalog.

CONCLUSION

In Search of Imagination

Genetics is offering us a way of reading the past, of reading history, in sweeps of time broader than we can imagine. Genetics is teaching us to read the connections, the ties, the relationships between all the forms of life on earth. We share genetic structures not just with chimpanzees or even dogs and cows, but with such simple life forms as yeast. I knead the bread and the cells in my hand and the cells in the yeast are written in the same language, with even, here and there, a recognizable fragment of a word, a DNA sequence, in common. But while some of us are left staring into the bread bowl pondering the wondrousness of life, others among us are running around the planet looking for "uniqueness," that which marks each individual and that which separates the various peoples of the earth from one another.

As science, the Human Genome Project, this attempt to map all of the genes, and the new genetic determinist thinking that has been accompanying it, is a lot like the man-on-the-moon project of a generation ago. It pulls together a lot of the scientific community, gives it a finite goal, and inspires much talk about human control. But as I think about the Human Genome Project, the words of a Leonard Cohen song keep running through my mind: "No, they'll never, they'll never reach the moon now. At least not the one we were after."

They wanted to understand our place in the cosmos. They found some interesting rocks. They want to understand the meaning of life. They're finding some interesting proteins. We're *always* trying to find the meaning of life and our place in the cosmos. The meaning of life is no more to be found in the genetic code than in the composition of rocks.

The problem is that genetic thinking is reductionist: it breaks things up, reduces them to the smallest possible parts—adenine, thymine, cytosine and guanine, the ATCG of the genetic code. Those letters are seen as forming words, creating a code for the construction of life. But the map of life they're drawing is without context: it stands alone. "Shy" genes don't blush; gay genes don't go off to gay bars; smart genes don't do math. These "genes for" have no meaning without a context: the context is the person and the intersecting communities in which that person lives. When I think of this map without a context, this environmentless listing of code, I want to write in the borders, "Beyond here there be dragons."

The dragons, some friendly and some not, are what they still cannot imagine: that which gives place and meaning to the map of the individual. People are more than their cells, and to understand that we have to look within the person to what we sometimes call the "soul," and we have to look beyond the person to the community, to the social world in which that soul comes into being.

What genetic thinking—what all reductionist thinking—lacks is an imagination, the leap of the mind that takes us beyond the pieces to see the whole. And that may be as close as we're going to get to understanding the meaning of life: the whole is greater than the sum of its parts. That is what defines life. Something is there, something that is more than the constitutive molecules, and that something is life. The "problem" of life, the thing that needs explaining, is, what is that "something"? In the language of physics, the problem of life is that it defies the second law of thermodynamics: life is not subject to the law of entropy. It does not disolve into nothingness, but *lives.* Erwin Schroedinger was the "father of quantum mechanics," a physicist and a philosopher, and someone who predicted the essential structure of DNA, what it had to be in order to work. Evelyn Fox Keller tells us how in 1943, as an older man, struggling to keep himself going while in exile from his German homeland, he answered the question "When is a piece of matter said to be alive?" "When it goes on 'doing something,' moving, exchanging material with its environment, and so forth, and that for a much longer period than we would expect an inanimate piece of matter to 'keep going' under similar circumstances."

Molecules themselves are just molecules: a living being *does* something. Not only does it keep on "doing something," an activity Fox Keller reminds us that an older man would surely value, but—with the eyes of a mother—life makes more of itself, it creates and recreates itself.

Trying to reduce life to DNA confronts this problem: DNA is not "alive." It is inert. It just sits there. DNA doesn't create life: life creates DNA. Robert Pollack, the biologist who told us that "children are assembled as a collection of discrete, randomly assorted, stable, dominant and recessive ancestral alleles," tells us that life is just DNA's way of making more DNA, something he calls a depressing, reductionist summary of natural selection. "To those of us who choose to make it so, life is more than that, but we cannot call on biological justifications for our choice." But is that all life is, even biologically?

We can get caught in this circle forever: does life make DNA or does DNA make life? What we must remember is that life itself is not DNA. It is not reducible to DNA. Life, whatever it is and whatever it means, is something more than the sum of its parts.

That is true at the level of the cell; true at the level of the person; true at the level of the society. Reductionism does not work: the cell isn't just a bunch or even a code of molecules; the person isn't just a mass or even an organization of cells; and the society isn't just a collection of people. In the organization itself, in the whole, exists something that was not there in the parts.

The map that they're creating is supposed to be a map of the person, blueprints for the individual. It is a map that represents the individual as DNA, as code. The metaphors are well and truly mixed: maps, blueprints, codes. What they share is their reductionism: they take the whole, the person, and reduce it to adenine, thymine, cytosine and guanine, arranged in base pairs, twisting up a spiral staircase, trying to find life.

The genes do not just unfold themselves, produce proteins,

grow into a body, become a person, organize a social world. The social world does not unfold from the genes, life as a product of gene action. Like a bread, life happens *somewhere,* and if it doesn't happen somewhere, some specific place, in its own history and place and time, it just doesn't happen at all.

The ability to speak is surely "genetic." But, just as surely, genes do not produce speech. Once upon a time, some kings were arguing about which kingdom had the real, true, natural human language. They could see that there were similarities in their languages, and they argued about which was the original, and which were the corruptions of the true tongue. So they raised a baby in total isolation from speech: no one ever said a word to that child. And they waited to hear what language the baby would speak.

Every way of thinking about the social world has, implicitly, a way of thinking about the individual; and every way of thinking about the individual has an implicit way of thinking about the social world. The kings thought of the individual as something preformed, complete, divine, and so they thought of society as something that could corrupt. Since nothing more could be added, society could only take away from the divine perfection of the individual person. The geneticists start with the letters A, T, C, G and build up to the social world. Genetic thinking, because it sees the individual person as dictated by genes, sees the society as an inevitable playing out of "human nature." If there is war, there must be "aggression genes" to account for it; if there's great disparity in wealth, there must be great disparity in genetic attributes, and so on. The social order is explained by looking at the characteristics of individuals, characteristics themselves explained by genes.

It sounds like a natural alliance with the political right: arguing from human nature to the social order is a conservative stance. But in its early days, genetic thinking as it played out in the eugenics movement attracted some very progressive, left-wing thinkiners. Genetic thinking does hold out hope for making a better world: you just have to make better people. The older eugenics involved clumsy attempts to control "breeding," encouraging more children among the better types, and discouraging, one way or the other, children of the lesser types. Contemporary approaches offer a far more sophisticated technology. But both are hopelessly mired in the reductionist thinking that fails to imagine the whole as something more.

This isn't the first time that sociology confronted reductionism and determinism and tried to point a way out. The philosophy of George Herbert Mead, the underpinning of the symbolic interactionist approach to sociology, grew in response to the reductionist and determinist interpretations of Freud and of Marx, and out of the biological determinism of the 1920s and 1930s. Mead put the person back in, the person as an actor, a person, we might say, with a soul.

Rather than reducing the person to a set of structures—genes, or ids and egos, or IQ, or attitudes, or any other set of things—the interactionists view the person as a process. Mind is a process by which people deal with the world, an ongoing activity. Judgments, perceptions, concepts, ideas—these are actions we engage in, things we do. We organize the world we live in, we make sense of it, we think about it. We're not just Pavlov's dog: between the stimulus and the response is a mind.

The self is a product of the mind, one of the things we have ideas about. But the self—who I feel I am, who you feel you are—is more than just a product; it too is a process. I am a process, I am what I am doing, but I am also something more. I see that, as we all see that in ourselves. We know ourselves, we see ourselves, we recognize ourselves. And we know, see and recognize others, and ourselves in them. We see others as people like us, and in interaction, we create a social world, an ongoing process. We exist in each other's minds, we grow and come to being in social life, communally, and our minds are in a sense fragments of the larger social mind: each mind is just a refraction, a prism, through which one sees the mind, the consciousness, the social world. We construct the world, and the world constructs us, in endless, constant process.

The self, the person, the being—who you are and who I am—is not just the location where determinist forces intersect. We are, each of us, more than that, more than the sum of those parts. Sometimes we talk about that process of acquiring a sense of self, of becoming our own true selves, as if it happens more or less on its own: "The child develops." I think about that process as a sociologist, but also as a mother. Children do not just develop. A baby not nurtured dies. Even if fed and kept warm, if it is not nurtured, if it is not mothered, it *dies*. The individual without a context, the blueprint sitting there by itself, cannot become, cannot be. Life ends, and entropy does take over.

This is true not only at the level of the social world, but at the level of the body, at the very level of the cell. We don't start out as separate beings. We begin as, we are in our very essence, social beings. In the relationship between the nucleus and the cytoplasm of the original zygote, as in the relationship between that cell and the womb, we are social beings—an interaction patriarchal thinking has made it very difficult to see. The cell itself, and fertilization, is seen with a gendered eye, as Fox Keller explains:

> The female gamete, the egg, is vastly larger than the male gamete, the sperm. The difference is the cytoplasm, deriving from the maternal parent (a no-man's land indeed); by contrast, the sperm cell is almost pure nucleus. It is thus hardly surprising to find that in the contentional discourse about nucleus and cytoplasm, cytoplasm is routinely taken to be synonymous with egg. Furthermore—by an all too familiar twist of logic—the nucleus was often taken as a stand-in for sperm.... Thus many debates about the relative importance of nucleus and cytoplasm in inheritance inevitably reflect older debates about the relative importance (or activity) of maternal and paternal contributions to reproduction, where the overwhelming historical tendency has been to attribute activity and motive force to the male contribution while relegating the female contribution to the role of passive, facilitating environment.

Of course the way that the scientists thought shaped the kind of research they did—and did not—conduct. Fox Keller discusses the significance of the studies of cytoplasm, studies that nobody bothered to do until the 1970s, not because they weren't technically possible, but because of "the belief... that the genetic message of the zy-gote 'produces' the organism, that the cytoplasm is merely a passive substrate." In other words, since it wasn't about the genes in the nu-cleus, it was "just" the environment, and no one cared.

The self, the essence, the power and action, came to lie in the nu-cleus of the cell, and genetic thinking took over. The whole rest of the world, from cytoplasm to cosmos, came to be thought of as the passive environment, at best the warm and welcoming place in which the action happens, at worst the dangerous, devouring, thwarting, obstructing mother. But it is not like that: the cytoplasm, that which surrounds the nucleus, isn't a passive background. What Fox Keller points out is that studies of the cytoplasm established "the critical role played by the cytoplasmic structure of the egg be-fore fertilization—before time zero." There *is* no time zero, no zy-gotic moment when the person starts. Johnny Jack's mother was right when she told him there is no *before:* "You were always in me."

What then is a person? When does it start? If it isn't the object writ by DNA, there in the plans, then what is it? At this time the question of personhood has been so taken over by the abortion de-bate that we are hard put to see it in any other terms. But the ques-tion of what makes a person is one that every society wrestles with, and in so doing, defines itself.

One of the ways of talking about personhood is to talk about en-soulment. Genetic thinking has so become the coin of the realm that it now seems almost self-evident that conception, the moment when genetic identity is laid down, would be the moment of ensoul-ment. But people had ideas about the soul long before they had ideas about conception as we now understand it, conception as the joining of egg and sperm to produce the genetic code.

In patriarchal thinking, the life force resides in sperm; it is planted in the body of the woman. But semen isn't people: it's hard to imagine baptisms or last rites for seed spilled, however innocently or guiltily. So when did the seed become a person? For the Catholic Church, from the beginning of the twelfth century when the ques-tion came up, through the nineteenth century, a distinction was made between an "unformed embryo" and a person. The journey of the soul to the fetus, a process they called "hominization," did not occur until 40 days after conception for a male fetus, and 80 days for a female. There was, actually, some question about whether the female ever acquired a soul: the Council of Trent debated the existence of women's souls in 1545, finally tabling the issue.

In the 1860s—the same time that Virchow discovered the cell, that Gregor Mendel observed genetics in action, when the American Civil War was being fought without a position on slavery being taken by the Catholic Church, a time when in many ways the understanding of the person was changing—the Church decided that hominization was not to be the basis of decision-making in abortion. With conception, they declared, there was a potential human being, and that potential was itself to be valued as a person—the beginning of genetic thinking.

Other traditions have placed that moment of personhood at "quickening," the moment when the baby makes itself felt. The word "quicken" refers to life, but it also refers to the first time the mother feels the baby move inside of her. Life is an act of communication. The baby reaches out and touches the mother, makes her feel its presence, and the mother gives the world a person. She declares that a new human being is among us.

Still others have waited for birth, for the appearance of the baby before the eyes of the community. Some use the moment of birth itself, and some opt for an occasion a few days later, to offer a ritual welcoming to those babies that made the passage safely.

And at the end of life too, we mark the passing of a person, and in so doing, recognize what constitutes a person. The anthropologists use the presence of ritual burial sites as proof of the humanness of a group: humans mark death. And in marking death, what we are marking is life, a particular life. Something is gone from the world when a person dies, and the rituals of death mark that passing, acknowledging as well the life that was.

Other advances in biotechnology, not just genetics, have forced on us this question of what makes a person. The discussion of brain death was about just this: at what point does a body stop being a person and become "the remains," just a body? How much *life* does there have to be for personhood?

And at the intersection of life and death, a discussion has opened up about the uses of anencephalic babies as organ donors for other babies. In this context we must ask, is it a baby the woman has given birth to, or the body of a baby? If it is a baby, then it has to be cared for as a baby for as long as it lives. If it is not a baby, not a human person, but only the body of one, then not only do we owe it none of the loving nurturance we owe a baby, even a doomed baby, but we are free to look at that body as we look at other bodies: as a thing, and sometimes even as a potential resource.

The person is never just a body, or just its actions, or just the time and space it takes up in the world: there is always that something else, that indefinable something that is life, that makes the whole greater than the sum of its parts. That person, that self, is not just the working of the genes, or the location where various determinist forces intersect. The whole is something more. Because every vision of the social world rests on an understanding of the self, of personhood, we need a way of thinking about the person that lets us create a good world. What can we call that something more, how shall we imagine it?

What about the idea of a soul? I posed that as a problem for my study group. We began working together years back on our dissertations, and people have come and gone over the years, but still we meet every few weeks, and over someone's table we talk about our work. What, I asked, can I say about the question of the soul as a way of thinking about personhood? Susan Farrell, who does research on women in the Catholic Church and who has training as a theologian, was troubled by my using the concept: "An awful lot of evil has been done in the world using the idea of the soul—separating soul from body, deciding who has and who has not got a soul, controlling the soul." Carolle Charles, who works on race, class and gender in identity issues, agreed, but added, "If that is all we have to counter the forces of the market which say that everybody is just a resource, grist for the postindustrialist mill, then go for the soul—it gives us something to hang onto. Marx said capitalism would ultimately leave no bond between people but naked self-interest: maybe the idea of the soul, the idea of something of innate and nonmarket value in each person, offers us some protection against that."

Judith Lorber, founding editor of the journal *Gender and Society*, and a strong social constructionist, reminded me of the way the problem has been handled in sociology, the way that Mead and the symbolic interactionists thought about it: there are various social "me's"—me as mother, me as friend, me as colleague—but in there somewhere, there's a sense of an *I*, an experience of myself as a person, an actor. A soul?

Eileen Moran, our most politically active member, said, "Over and over again throughout human history, we invent an essence, an inner true self. I'm not troubled by that, that eons of human history have invented the concept of the soul to say that there is something precious about individual human beings. We still need that, even if science, and some social science too, tries to court us into another perception."

They're right. All of them. I know, as Susan does, that there is no idea that cannot be used against people, used to divide and harm and separate. But Carolle's pragmatism is also right: if the idea of the soul offers us some moral authority with which to counter genetic thinking, then let's use it. The soul—the self, the precious something that makes each individual life worthwhile—is something that we have a feel for in our individualistic culture. We can use that as a protection against the cheapening of life, when genetic thinking reduces life to a list of instructions, randomly assorted alleles. And like Judith and Eileen, we must recognize that the soul is socially constructed, in both meanings of that term. The soul is itself a creation of the social world, made and formed and brought into being in a social way, between people. And our ideas about the soul, what it is and what it means, are also social ideas, ideas we share.

The leap of imagination we need is to see that that soul, that self, that individual, is made not just from the letters of a code, but also from other people. That which is holy and precious and unique about each of us is also that which is shared between all of us.

Existential Orphanhood and Identity Politics: On Jews, Dwarfs and Gays

Every community is organized around the question of what constitutes a person. That question is asked at the borders of life, at birth and at death, but it also is asked at the borders between people. Are women really people, full people with full souls? Are "heathens"? Are animals? Which animals? Why or why not? Are they—the other, whoever and whatever the other is—are they really us?

We can answer the question of personhood more generously, more inclusively, or we can answer it more narrowly and exclusively, but we always have to answer it when we form a community. I have a poster hanging on my office wall from the Fortune Society, an organization of ex-convicts. It's been there so long I can barely make out the fading words: "Never build a wall until you know what you're walling in and what you're walling out." It's good advice, coming from people who have been walled in—and so walled out.

How do we form communities? We talk about "identity politics," but that is never about individual identity, who *I* am; it's always about group identity, who *we* are, and how individuals take their identity, their sense of self, from communities. I value community. The world that I live in, and the world that I want for my children, is not a world of scattered isolated individuals, and not a world of walls. It is a world of communities, of social solidarity, of connectedness between individuals and between communities, a world in which people and communities grow from and into each other.

Why, when we build community, do we so often choose just those things we think of as "genetic" as the basis of our community, as the basis for our "identity politics"? Why do we so often choose race in particular as a way of placing ourselves and others inside— and so outside—of communities?

We don't. Race chooses us.

That assortment of bits and pieces, lines of A, T, C, G play themselves out, and the angle of a nose, the production of melanin, the shape of the hair shaft, writes race upon the body. A story is written into our bones and skin and hair, there to be read by anyone who knows the code.

I live near a school. Every year or so some little kid comes blubbering down the block, lost. No one picked her up. All the school busses, stroller-bearing parents, all the crowds have cleared, and this one last lone child has given up waiting. I'm one of the few people working at home on this largely empty street in the afternoons. I've brought kids in and given them snacks while we tracked down parents; left a more nervous kid on the front steps while I made calls; walked a kid home when he knew home was "over there," but not sure where. I was walking one such kid home, a scared little boy of about five. An older woman stopped us and went to take the child from me. "He's one of mine," she said. The kid didn't seem to know her, and I was confused. "One of yours?" "One of mine, one of my people's, one of ours." She began to speak to him in Haitian Creole; I was thanked for my time and off they went.

She saw his face and knew he was one of hers, her responsibility, not mine. One part of me resented that—hey, I'm a competent mama, I was handling it; one part was relieved, a time-consuming task taken away; one part just generally admiring and appreciative of the sense of community, of responsibility her actions displayed. It takes a village and all that.

When Victoria was little we had variations of that happen all the time. If she'd walk along even a few feet ahead of me—and that kid never walked, she hit the ground running, scampering, skipping, cartwheeling, twirling—people saw a lone little black kid and no parent with her. I was invisible, white, faded into the background. Some random black woman following behind her would have gotten the parent-to-parent smile over a cute kid's antics, but not me. If no black folk were immediately in her vicinity, well-meaning people, white and black, would do what I probably would have done: ask the kid if she's lost. "Where's your Mommy?" A moment's confusion, an embarrassed smile. I learned to repeat a line I'd heard from another white mama of a black kid: "It's okay—it's not obvious."

Race makes it obvious. Race writes community on the body, and constructs community for us, makes us readable and placeable to anyone who knows the code. But it's always about knowing the code. At the New Zealand Maori ceremony, I looked at that fair young man next to his darker uncle, and never ever would have read their relatedness. I didn't have a clue. A few more months in New Zealand and I too probably could have learned to recognize a Maori nose, read from that clue an "obvious" relationship that to me, at that time, sophisticated and well versed in another language, was hidden.

Race is a code written upon the body. Historically and consistently that code has been used as a weapon. People read race to exclude, to identify the other. Knowing the code let people pick up runaway "mulatto" slaves, catch Jews with false papers, identify the "half breed" moving into town.

Race is historically, consistently, a weapon. Can it be anything else?

It was something else in the hands of the Haitian woman who claimed the lost child from me. It was the basis of community. It was a fragile line in the web of community that surrounds and protects that child. Haitian boys need protection in Brooklyn. As I write this a big news story in New York is about the beating and torture of a Haitian man by the police. It's in my precinct.

Every time you draw a line, you create two sides. The cops drew a line between themselves and a Haitian man, Haitian men. The woman drew a line connecting herself and that child and put me on the other side. If the state, if the larger society, if America hadn't drawn that first race line, would that woman have drawn the second?

Could I have walked up to a black woman assisting a lost white child and taken the child, claiming it as one of "mine, ours, my people's"? Can I claim "whiteness" as a community—from which a black woman is excluded? Would I participate in any way in drawing the line that constructs whiteness? God no, not in this place and this time.

I carry with me the privileges of whiteness, written on my body. What is the whiteness I have? Eastern European Jewish whiteness? What a mass of contradictions right there, and yet my skin, with its genetic code to limit melanin production, having been brought to this continent takes its meaning from the race code here, not the one that worked in eastern Europe. I am what you read me to be.

I'm not much of a Jew, not very "identified" as a Jew. I was in Switzerland, in Zurich, on Kol Nidre, the night before Yom Kippur, the holiest of the Jewish holidays. I should have been home—if I was much of a Jew, I surely would have been—but I was speaking at a conference, and what with school schedules and all, it was my chance to see Switzerland. I should have been home—the raw bread dough for the traditional challah that I prepared and froze got left out too long and collapsed. Had I been home, I'd have cooked, lit holiday candles, served a filling dinner for whoever might observe the fast the next day, ushered my in-laws off to services. I would not have gone. I never go to synagogue. But I was in Zurich, and I was intrigued, and maybe something else too. I was in Zurich, where German was being spoken all around me, where the gold of Jewish teeth probably still sits in vaults somewhere. I took a city tour, asked where an active synagogue was, and walked over there in the afternoon to check what time services were.

As I walked up to the synagogue, a young man walked up. I forget where he was from—somewhere in Asia he said, though he sure didn't look Asian. He was working in Zurich for the week and was there on the same mission I was. *"Gut Yontif,"* we wished each other— "happy holiday." I'm not even sure if that cheery a greeting is appropriate for such a sombre holiday, but we offered it to each other. We, two strangers, two Jews, tossing a few words of Yiddish into the German-heavy air.

I'm not much of a Jew—but I am *that* much of a Jew. That stranger from another part of the world and I, meeting by chance on a Swiss street corner, tossed each other a line, a thread, the slender strand along which community is constructed.

We need to be part of a community, connected to others. We, us, we who understand, talk the language, get the joke, share the grief, catch the knowing glance. "We"—a wonderful word, a terrific feeling. The other side of what makes I, me, myself a possibility: the "we" from whom "I" take meaning, strength, solace, comfort. The "we" who enable "me" to go out there in the world, to face "them."

It's not only about race.

A woman physician friend of mine, 20 something years ago, had to call in a prescription for a patient. Pharmacists always assumed that she was the nurse, the assistant. She got a woman on the phone at the drugstore, explained that *she* was the doctor. The woman explained that *she* was the pharmacist. They laughed: "We are everywhere!" When she told me, I laughed. We *are* everywhere.

I spent the better—well, actually, the worse—part of a year on crutches with my broken ankle. At school the elevator button is in the middle of the elevator bank, and there was no way I could maneuver to the elevators on the far side before the door would slide closed. By the time I hobbled to the left, that one was gone and the one on the far right was arriving. I called out one afternoon, in some desperation, "Please hold it for me—I'm on crutches." I got there—and the man moved back with his seeing eye dog. We laughed. We are, God help us, trying to get everywhere.

I sat at dinner in Chinatown with my two daughters and my husband. A family came in and sat at a table across the way. Two white parents, a white grandma and two brown kids. Two to my one. We smiled at each other. Victoria said, look, they're like us. I smiled. We are everywhere.

My husband and I were driving somewhere in Vermont. Daniel, my oldest, my only kid at the time, was a baby, just a few weeks old, in the infant seat between us. We pulled over beside a field so I could nurse him. Breastfeeding was just coming back into vogue in America—I knew just one other woman who nursed a baby. I unhooked him from his straps and he suckled. A cow was standing not far away. A calf bounded up and suckled. I laughed. I don't know what the cow thought, but I thought, we are everywhere.

A Jew, a woman, a feminist, passingly disabled, a member of an interracial family, a nursing mother—I have a thousand identities. They're like hooks that stick out all over me, catching strangers and finding connections. Points between which lines can be drawn. Lines that build to a web, a net, a fabric.

The identities we have, and the communities we form, are often defined in terms of biology: communities of race, the gay and lesbian community, the feminist community, the disabled community— and by the very use of the word "community" we focus on the social rather than the biological ties that bring us together and keep us apart. The biology of it, the genetic code of it, isn't what unites us: we unite ourselves.

What each of those communities has in common is where they sit in the distribution of power. I couldn't imagine using my refound able-bodiedness as the basis of community, or my heterosexuality, any more than my whiteness. When people do that, when they try to form identity-based political movements from positions of power, it is usually because they feel, or claim to feel, threatened. Think about the rhetoric of the Nazi party, Germany in the post—World War I mercy of the world; of the Klan in the South after the Civil War; of some movements for "men's rights." They claim—and often genuinely and sometimes appropriately fear—loss of power, loss of privilege.

Identity politics is a rallying cry for people whose identity is imposed and devalued. We unite from below, and try to use the messages that have been written on our bodies not to separate, not as weapons against others, but as the threads from which we can weave a community.

I was at a "women and genetics" conference in Zanesville, Ohio—getting to see a bit more of the world, I suppose. Jacki Ann Clipsham, an artist who is an achondroplastic dwarf and uses a wheelchair, was presenting a history of disability in art. She showed slides of people with various kinds of disabilities as portrayed in painting, drawing and sculpture over the last several hundred years.

Sitting there in the darkened room, my hell-in-a-handbasket embroidery (which I now bring to all genetics conferences) quietly resting in my lap, I lost myself in her story. I listened to Jacki's history of disability, and was struck, moved, by her use of the word "we." It was the way she spoke of the dwarfs represented in those paintings: we were slaves in the royal courts of Italy; we were gifts to children, kept as glorified pets for princes; we were the jesters and entertainers. *We.* My mind drifted: Remember that you were brought forth from the land of Egypt, and speak of it in each generation as if you yourself were brought forth. It was the echo of the phrase from the Passover seder stirring in the back of my mind. Or was it the tribal nature of Judaism and of dwarfism that caught my attention, that caught and created that echo? We, of the tribe, must speak of our history in each generation, teach it to our children, and wear it as a sign that we may remember.

Jacki wears her sign upon her body. She sat there in Zanesville, Ohio, an artist from New Jersey, speaking the history of her people: in Italian courts, in the French countryside, in ancient China. They were her people, in diaspora. They were everywhere.

Achondroplastic dwarfism is a single gene condition: one mutation and the growth pattern changes dramatically, and people come to have that recognizable shape. It's a body shape associated with childhood, even babyhood: the large head, small limbs—in a way, rather endearing. But it is a shape not generally associated with adult sexual development, and so presents a sometimes comic juxtaposition with beards or breasts, other signs of adulthood and sexuality. Probably not an easy body to live with, surely a tough adolescence: growing up without growing *up.*

That gene constructs a shared experience; that experience becomes the basis for constructing a community, a *we* across time and space, a people, a tribe in diaspora. I came home and read about dwarfs. I saw I wasn't the first person to have made one particular connection. Leslie Fiedler, literary critic, called the dwarfs the "Jews of the Freaks": "the most favored, the most successful, the most conspicuous and articulate; but by the same token the most feared and reviled, not only in gossip and the popular press, but in enduring works of art, the Great Books and Great Paintings of the West. They have been, in short, a 'Chosen People,' which is to say, a people with no choice."

Dwarfs have long captured the imagination, not only of painters but of writers too. Dwarfs have been featured in best-sellers twice recently, in John Irving's *A Son of the Circus*, and Ursula Hegi's *Stones from the River.* Hegi uses a dwarf woman, oddly enough, as the protagonist in her novel of life for Germans under the Nazis. What makes it odd is that Trudi, the dwarf, remains unthreatened by Hitler's genocidal actions against dwarfs, his targeting of dwarfs along with Jews and gypsies and other inferior stock. She lives quietly, strangely safe in her small town. The dwarf becomes Everywoman, the truly good German, her stature ensuring her enough of an outsider status so that there is never any question of her being one of the bad guys, but not outsider enough to be anything but German.

In writers' fantasies, dwarfs are a people, a magic race perhaps. Fiedler quotes Pär Lagerkvist's novel *The Dwarf:* "We have no need to be fertile, for the human race produces its own dwarfs. Our race is perpetuated through them."

I think of Linnaeus, identifying the four color-coded continent-bounded races of the planet, and that fifth race, scattered in diaspora: monstrosus. Hegi presents it most eloquently, in a passage in which Trudi goes to a carnival and meets Pia, a dwarf animal trainer. Pia asks for volunteers from the audience for her act, and Trudi walks up: "For a moment Pia looked startled, and her black eyes skipped past Trudi and back as if snared by her own reflection. But then she laughed with delight. 'Come.' She extended her free hand and Trudi held herself straight as she walked toward Pia. 'It looks like we have a volunteer. From the magic island which I call home. The island of the little people, where everyone is our height.' " Between them, they describe the magical island, calling it forth as if from memory. Afterwards, Trudi goes back to Pia's trailer. "There must be others," Trudi blurts. "I have never met anyone like me." Pia assures her there are, everywhere. She has met 104. ("You count then." "How can I not?") "Dizzy with joy, Trudi could feel them—those one hundred and four—linked to her as if they were here in the trailer."

If my daughter Victoria's skin and hair and bones mark her as African, if the genes that shape her body mark her place in the race scheme of America, how different is that than Trudi's or Jacki's achondroplastic gene marking their membership in the world?

Race gets passed along, inherited in that old-fashioned sense, genes moving from parent to child, marking, making obvious, relationships. The dwarfism gene pops up, springs itself upon us here and there, cut loose of the moorings of family. Does it mark peoplehood, community, relationships in a fundamentally different way? Or are we using the resemblance that genes write upon the body as a way of relating, connecting, weaving people together, just as profoundly as we use them to tear people apart, to cleave lines between, separating rather than joining?

I haven't raised a dwarf child. I've been raising a black child, and I know I've needed the help and friendship and support of black friends to do that. I needed a black 16-year-old kid to spend an afternoon with me, with jars of cream and grease and oil, with combs and brushes and bands, to teach me how to do hair. And, being a slow learner, I needed to be taught that more than once, by more than one friend. I've needed black friends to help me sort paranoia from racism. I've needed black friends to just be there, to be the ordinary parts of our lives, to make blackness and whiteness variations in skin, not separations in the world.

I knew, when I offered to raise Victoria, that I would be raising a child in one world for another: that there are separate worlds of black and white in America, and that however my life tries to straddle and spread and blur the lines, the lines exist here, and it is a line she has to cross. The sociologist Heather Dalmage calls this crossing "tripping on the color line": tripping as in stumbling; and tripping as in playing, fantasizing, enjoying.

But truth be told, we raise all of our children for worlds we don't, won't, can't inhabit. With Victoria I could see it coming before I began. Daniel made sure I understood that lesson when he came out as gay at 14. My world crosses that line too—I have lots of lesbian friends (some of my best friends!), a few gay men friends, but my world is on the other side of that line, the other side of a line Daniel crosses. Is that line too written in genetics, a "gene for" gayness? Does it *matter?* Call it a gene, call it a combination of biology and circumstance, call it life itself playing out anyway it chooses: I'm here; he's there. I parent him in this world for that world, raise him to be his own true self in his own place, and that includes his place as a member of the gay community.

And so it is with all of our children, and so it is with all of us. We walk through different worlds, we move between. However it happens, we take this thing called identity, the qualities that make up ourselves, and we use them to construct relationships, communities, social worlds. It may start with a single gene that mutates like a wild card, a joker through the deck, dwarf children, and gay children and deaf children, children of all sorts of diaspora, springing into being here and there all over the world. Or it starts with an evolutionary lightening of skin or slanting of eye over time beyond time, and whole continents of people separate out into light and dark, this and that, us and them. It starts with one gene, or many, or none. Identity marks itself upon us, sometimes writing itself on our bodies for others to read, and we go from there.

Some of where we have gone has been very evil. We have used those markers to isolate, to destroy, to enslave, to kill, to torture, to maim: to hurt in every possible way. The exclusion is there, an absolute, unshakable part of our history. But out of that exclusion, time and time again, have been established communities: it is the good that rises up from the evil.

Quentin Crisp, who raised being homosexual from a condition to a lifestyle to an art form, says,

> When you're born a Jew, you have Jewish parents.
> When you're born a black, you have black parents.
> But when you're born gay, you're an orphan.

Orphanhood is our modern condition: the existential orphanhood of us all, without parents to guide us in the new worlds we inhabit. As the self splinters and crystallizes and scatters, we put the world back together by connecting the pieces between people. We draw the lines, from one to another, that remake the world, that unite, and tie and bind.

We toss those fragile lines across to each other, and establish community. In ghettos, in sideshows, in slave quarters, in schools for the deaf, in gay bars, in all kinds of nooks and crannies throughout the world, people have established community and connectedness. They weave a fabric, and use that fabric to catch and to hold the next generation. The black woman who keeps an eye out for the lost black child; the gay guidance counselor who reaches out to the troubled gay kid; the circus dwarf who answers the call of the frightened dwarf child who has "never met anyone like me,"—these people are all building community, creating a safe world within the profound limitations that the wider world may offer them.

It may—or it may not—start with the tangled ATCG of the gene, but it has moved to the language of a people. We, us, we who know, recognize each other, reach out, and build what we can. We are everywhere.

After I Finish

I want order and control and predictability in my life and in my children's lives. I don't want to live with chaos, and randomness, and not knowing what to expect next. When change comes, I want it to come peacefully, with time for me to prepare.

Faith Ringgold, an African American quilter and storyteller, said she had a good childhood: nothing happened.

That's the kind of childhood I've wanted for my kids, and I guess in a way that's what I want for me, too. When something happens, I'd like it to be because I wanted it to happen, because I was ready for it to happen.

It's not going to be like that. I'm going to face the loss of every single person I love and care about, through their deaths or through my own. If I surprise myself, as seems increasingly likely in my so-far-healthy middle age, and don't die young, I will face aging and the loss of parts and pieces of myself, my health and my youth. Anything I don't lose to death in the course of my life, I lose to time itself. I've lost all my babies into the children they became, their infancy as gone to me as ever gone can be. My little sister, I'm astonished to see, is a middle-aged woman almost! My baby brother is a no-longer-young man. Most of my aunts and uncles are gone. Everything and everyone goes.

We are not going to live forever. There are no happily ever afters. Life is a very scary place, and we're not getting out of here alive.

Any wonder people want to play God?

I can't really profess horror at people who want to play God. I remember a character in a novel pondering the organization of human life and snapping, "I could come up with a half-dozen better arrangements before breakfast!" Or a Yiddish expression: "If God lived here, He'd have His windows broken!" Life is not so perfect that it couldn't stand a bit of tampering.

The geneticists have taken an interesting position on this: they use the God-language—the "book of life," the "Holy Grail." God, who used to be a watchmaker, has been reskilled, and is now the Prime Programmer. But the code isn't just God's domain, they say: they don't seem uncomfortable with improving on the code, adjusting it, fixing it, enhancing it. At what point, one wonders, do they think they're creating new code, creating life? To play God—to *work* at God—is to have power over life itself, over which lives to call forth, to create. Poems and books and such are made by fools like me, but only God can make a tree. Sort of. Trees are cloned now quite regularly. Is that *making* a tree? It's copying a seed, but then, as far as I can tell, the seed makes the tree.

But the scientists do seem to think they're "creating" life these days, rearranging things, taking bits of DNA from here and putting it there, moving it from one species to another, copying at will. The new ability to move genes between species, even between life forms as diverse as fish and plants, or bacteria and humans, makes hybrids and grafting seem quaint. It does seem that something noteworthy has been done when a new life form is put together. Constructed. Developed. Created?

OncoMouse, Donna Haraway reminds us, is our sister, a mouse with our human cancer genes inserted to order. OncoMouse, whether she was constructed, developed or created, has most assuredly been patented: she is the first patented animal in the world. For $50 to $75 a mouse, researchers can order these mice with transplanted genes. Haraway cites the president of one of the biotech firms that sells such creatures as saying that the technique of custom-making rodents is so routine that he calls it "dial-a-mouse." It's not only the human genome that is being mapped: the mouse genome is also hotly pursued.

I'm not sure how much sisterly love I can muster for Onco-Mouse, but I did feel enough kinship with cows to stop eating them when I learned how all the technology they were developing for human in vitro fertilization and related technologies of procreation was coming from, and going back to, the cows. That forced me to think about where I draw the border, the wall that separates *me* from *them*. How do I decide which lives—including animal lives—are in and of themselves to be valued, and which lives—including yeast and grains of wheat—are just resources for my use?

That is the question we ask about destroying life: what (or who) can I kill and still be a moral person? But it is also the question the geneticists are making us confront in creating life. What, or who, can we create, and for what purpose? I think about that as a mother. It's something I know something about, making lives.

There is a sense, when you are raising a child, that you are "calling it forth." It is both there, and not there. You are engaged in a creation, an act of creativity, perhaps *the* act of creativity, but there is another person there, the very person you feel yourself creating. It is part of the experience of pregnancy, of feeling yourself become self *and* other, but it is also a powerful part of the experience of looking into a newborn's eyes and calling forth the being that both awaits and has yet to come into existence. What are you thinking, we wonder, as we stare into those fresh eyes. Without language, without words, without experience of the world, what thoughts can be there? What thoughts can there *be?* And as we give the child language and words, we both create and call forth the thoughts and the thinker.

When I was working on *The Tentative Pregnancy*, my book about women's experiences with prenatal diagnosis, I reached a kind of despair. Leah was a baby, and the kind of baby that moves along slowly, at her own pace. She didn't learn to crawl until she was about a year old, didn't learn to walk until almost a year and a half, hadn't said a word until she was two. Once she was with me at a friend's party, and an elderly grandmother said to me, "How old is the baby?" I told her, and she said, "Well, don't worry."

She was that kind of a baby. But, oddly enough, I wasn't worried. That wasn't where my despair lay. I was entirely sure that there was indeed a responsive person in there, and that beautiful sweet face held understanding, and those pudgy little legs would get moving in their own sweet time. As indeed they did, and all is well. But at that same time I was mothering that baby, I was interviewing women who had terminated pregnancies because they were told that their babies would be retarded, would be slow, wouldn't develop "on schedule," and that ultimately all would not be well with their babies. And as I looked at Leah, I kept trying to figure out was it that I loved her as passionately, as intensely, as totally as I did because I knew she would be okay, or did I know that she would be okay, that "it" would be okay—her life, our lives together as a family—because I loved her so? The midwives talk about that, how a woman will say when she learns her baby is not all right, "It doesn't matter anymore. She is in my heart and I love her." It's not what they wanted for their child, but it is what they have for their child, and it is what it is. Life is what it is, and you go on, and you love.

To question that is to question the nature of mother love, of all love. What is the nature of this love? What is the nature of faith? What is the relationship between love in the present and faith in the future?

It is troubling as abstract questioning. It is incapacitating while you are doing the acts of love and faith that make it okay if it is ever to be okay.

And therein lay my despair.

To mother—or to deal with other people in the deeply nurturing way that is motherhood—we have to start from the position that the other is there. We assume a self in the other person, and in so assuming, call it forth. We start with the belief, the value, the commitment that the other is a person of intrinsic wholeness and worth: we posit a soul.

It's not a bad way to approach people.

Michael Bérubé writes about raising his son Jamie, who has Down syndrome. He shares the joy that Jamie experiences in understanding and in being understood—and in understanding that he is understood. The joys of communication: mutual realization and the reciprocity of life. "Communication is itself self-replicating,"

Bérubé tells us. "Sign unto other as you'd have them sign unto you. Pass it on."

And passing it on is what we do when we nurture and call forth the other. It's what Bérubé does with Jamie. It's what I did, more easily, with Leah. We act toward the child—we act toward each other—with love and attention and respect and expectation. Jamie may never do long division, never bring home an A in science, but Jamie is a person, Bérubé reminds us, a person with whom we engage in reciprocal communication. The only way to bring children and all of the people of the world to their fullest intelligence is to nurture them as if that intelligence is there: to posit its existence and in so doing, call it forth. Intelligence, goodness, humor, compassion—the very self itself. We have to take these as given to give them life, believe in their presence to call them forth.

That may be what each loving parent does. But that is not what we are doing as a community, as a society. As individuals we may approach each other with respect and appreciation, but as a society we approach each individual as a potential resource, a potential contributor to or drain on the economy.

People map what they value, what they think is important. We are choosing to map the human genome. A map is a guide: it tells you what is there, but it also tells you how to get places. The genome map, as we have seen, may end up taking us places we don't want to go. What with slippery slopes, camels at the tent, and the way one thing leads to another, the world that we're creating with this map may not be a world in which we'd want to live. But how can anyone argue against mapping, against exploration? How can we stand here, of all places, in the "new world," and wring our hands and fret about exploration? Yet maybe exactly here, the land that native people lost to explorers, the land where the language blends the words of the British colonizers with the rhythms of the African slaves, maybe here in this unmeltable pot, we're best placed to think about maps, where they take us and where they lead.

Why are we spending our precious dollars on this map? What are we looking for? If our concern is with preventing disability and disease, with extending the life span, with lowering infant mortality, with making smarter people, there are better places to spend money. They want to find the genetic contribution to intelligence? Appreciate each baby with wonder and awe at its developing mind, nurture that intelligence in a rich and stimulating and safe environment, offer schools that inspire as well as instruct, and then get back to me on the question of individual—or racial—variation in intellectual potential. Don't start with schools that warehouse kids, televisions that stupefy them, and parents who are overburdened and overstressed, and then look for the genes for intelligence.

Cars, violence, lead and other poisons in the air and water, all kinds of things are destroying our children, their bodies and their minds, at higher rates than our genes. A mother's zip code is still the best predictor of infant mortality. Deal with poverty, and then get back to me on genetics research.

Do we have enough vision, enough of an imagination, so that something other than the market might shape the development of genetic technology? We start out up at the top of the slope, concerned with genetic conditions that cause pain and suffering, and we try to make things better. With no market inducement, with no one standing to gain or to lose economically, I think we could look at a gene that appears to cause a baby to suffer and die in agony, that wreaks havoc on the body of a young man or woman, that destroys the mind of a person before the fullness of a life span, and say that gene needs to go.

But down here at the bottom of the slope is the market. Which kids would be cheaper to raise? Which would be the most efficient workers for the current economy? We can find ourselves in the equivalent of the square-tomato project, trying to engineer life for a purpose, for a particular need that is market-created.

We don't have a way of protecting ourselves from that market-driven need. We try to protect individuals from the kind of coercion presented by the old eugenics by giving them free choice, but "free choice" is never free. It's a choice made in a context, and not a context of one's own choosing. There is a deep unwillingness to think collectively in America, and it limits our imagination in profound ways. We see each individual as an individual, and give that individual what freedom we can. One at a time.

If we have no social use, no economic place for a "certain kind" of person, we don't have to legislate against those people: we just give each parent the individual choice of bringing, or not bringing, such a person into the world. Individual choice doesn't stop social engineering. It's a mechanism for achieving it.

We value the individual, we value personal choice, freedom for each and every person. It is America's great strength. But what is a person? We've spent the past several hundred years expanding our definition, making it more and more inclusive, extending personhood and citizenship to larger and larger categories of people. But now the person itself begins to fragment, crystallize into 100,000 separate instructions. "This is you," the geneticist says of the compact disc listing the code, "This is you." What constitutes identity, the sameness of self that transcends time and change? I am me—from childhood through all the changes of middle age and maybe onto old age; with scars and bruises and losses, my identity as self holds. Wherein lies that sameness, that identity, that self? You want me to believe it lies *there*, in a list of instructions for the formation of proteins?

For some people that essential self resides in a divine soul: there is an essence, a constant and irreplaceable soul that resides in, but is not of, the body. I use the word, I posit a "soul," but I cannot mean it in that historical, theological sense. And in our secular culture, in our developing genetic thinking, where has that soul gone? It has moved into the genome. In this genetic thinking, the DNA is where we've placed the core of the self and the essence of relatedness—into the nucleus, the program, the blueprint, the "genes." And we map.

The mapping of the European explorers changed the world. It changed the way people saw the world and it changed the way people used and were used by the world. What will this new map do for us? What does this genetic thinking tell us about who and what we are, what life itself is? If our essence isn't a solid core, an essential or divine soul, and if it isn't a liquid that flows from us to our children, then what is it? It crystallizes, it fragments, and so it scatters. There is, when you come right down to it, no there there. The center is not holding. If you believe the geneticists, it is—we are—just 100,000 pieces.

But that, I know, is not true. There is, in me and in each of us, in the children we bring forth and call forth, a center, a solid being within. It doesn't lie in the nucleus of a cell, and it doesn't unravel into 100,000 interchangeable, infinitely manipulable bits.

Where can I place that essential self? If I don't have the comfort of believing in an everlasting soul to fall back on, and if I don't place the self in the nucleus of a cell, then where am *I?* Where is the essence of a being, where is the child I call forth coming *from?*

I wouldn't mind knowing where the self really is located. My husband says "two inches behind the eyes—isn't that where everybody feels it?" Actually not. Higher maybe, more where I get a headache. Less vision-dependent for us myopic types.

But seriously, if you cannot, as I cannot, accept the idea of a material-less immortal soul, then where do you place the self, the true core, the center of being, who you *are?*

I spoke to Ruth Hubbard. I'm about done with the book, I told her. I still haven't figured out the meaning of life. I can't even find where I am. Where *is* the self, Ruth? Who are we and what are we doing here?

"Look in the mirror," she said, "There's the self."

I sighed. "Yeah, you're right, but ..."

"And smile," she said.

I finished the book. Why am I who I am, you who you are? And what is the meaning of life, the point of it all?

I looked in the mirror.

Smiled.

And winked.

Why am I me and you you?

Because the chicken told a lie.

Acknowledgments

Books are also products of interlocking communities. Many, many people provided the nurturance that made this book possible. Some of them wander through the book like characters in a novel or play; others worked backstage.

I want to thank them all:

The members of my study group, Judith Lorber, Susan Farrell, Carolle Charles, Maren Lockwood Carden (who was also my New Zealand travel companion and added enormously to my appreciation of that experience), and Eileen Moran, who not only is always on the same page with me, but often knows what chapter that page should be in.

Several people read and offered helpful comments on various portions of the manuscript: my thanks to Gail Garfield, Ruth Ricker, Jacki Ann Clipsham, Ira Van Keulen, Nora Jacobson, Setsuko Matsunaga Nishi, and Mary Beth Caschetta—who also gave me the working title "Of Maps and Imaginations," and put in a very crucial hour at an early and desperate moment.

Various people provided an assortment of research assistance on this project, including Janet Gallagher; Stanley Jin, whose services were made available through Baruch College; and Tania Levey, thanks to Electa Aranel of the Women's Studies Program at the CUNY Graduate School. My mother, Marcia Katz Berken, answered quite a few odd questions as they arose, and even made one urgent trip to the library for a missing book: for that, for many hours of child-care assistance of late, and for lots else besides, I thank her.

The Society for the Study of Social Problems granted me the great honor and fun of serving as its president, and of organizing the annual meeting of 1994, giving me an opportunity to explore these issues. I thank

Tom Hood and Michelle Smith-Koonz for their help, and particularly Alan Spector and Patricia Ann R. Flynn of the program committee for helping to frame the issues.

The Council for the International Exchange of Scholars provided the Fulbright award; Tjeerd Tymstra and his family made it work. The PSC-CUNY award of the City University of New York provided additional funds for the research in the Netherlands and for other work on this book.

My agent, Carol Mann, has always been a pleasure to work with, offering the kind of sensible, reaffirming advice every author should have. Mary Cunnane was my editor at Norton for a very long time, and I appreciated her help in the early stages of this project. Alane Mason, the editor who saw it through, was wonderful at pointing to various gaping holes I'd overlooked.

Ruth Hubbard saved me from making a fool of myself over various bits of detail, but much more than that, I value her judgment and her friendship.

Heather Dalmage got this project back up off the ground at a difficult moment, and has consistently offered helpful and insightful comments. She and Wendy Simonds have been there for me in a way that is very precious: with love, with support, with great intelligence and with friendship.

My family has wandered all through this book, so anyone reading it knows how much I owe them. My husband, Herschel M. Rothman, has been supportive of every project I've ever taken on, and has made my academic and scholarly life possible with that support. If my sister, Linda Katz, hadn't given me a passport and shown me the world, just think how much more boring this book, and my life, would be. My children have given me the world in other ways, continually opening up new worlds to me. It is a particular pleasure to thank my son, Daniel Colb Rothman, for helpful, thoughtful critiques and conversations as this manuscript progressed. For me, he and my daughters, Leah Colb Rothman and Victoria Colb Rothman, *are* the meaning of life.

Notes

INTRODUCTION

Before I Start

The phrase "prism of heritability" is from Troy Duster, *Backdoor to Eugenics* (New York: Routledge, 1990), and the phrase "discourse of gene action" and the quote in this section are from page 4 of Evelyn Fox Keller, *Refiguring Life: Metaphors of Twentieth-Century Biology* (New York: Columbia University Press, 1995). Dorothy Nelkin and M. Susan Lindee have written about the gene as an icon in *The DNA Mystique: The Gene as Cultural Icon* (New York: W. H. Freeman, 1995).

The book by Pearl Buck is *Johnny Jack and His Beginnings* (New York: John Day, 1954).

I have been influenced by the work and the friendship of Ruth Hubbard for many years now. Hubbard's point is drawn from page 7 of Ruth Hubbard and Elijah Wald, *Exploding the Gene Myth* (Boston: Beacon, 1993).

Alan Spector made the point that the soul has moved to the gene, and many other points that encouraged me in writing this book, in his 1992 presentation to the Society for the Study of Social Problems, and in his work on the 1995 annual meetings of that organization. Part of this chapter is drawn from my presidential address of that year, later published as "Of Maps and Imaginations: Sociology Confronts the Genome," *Social Problems* 42:1 (February 1995), pp. 1—10.

Donna Lee King gave me a copy of Dr. Seuss's *The Lorax* (New York: Random House, 1971) and is the author of *Doing Their Share to Save the Planet: Children and the Environmental Crisis* (New Brunswick, N.J.: Rutgers University Press, 1995).

On Breads, Bibles and Blueprints

For a very readable and far more comprehensive explanation of genetic science, see Ruth Hubbard and Elijah Wald, *Exploding the Gene Myth*. Another good introduction to genetics for the nonscientist is offered by Chandler Burr, "Genetic Grammar 101: A Crash Course," chapter 6 of *A Separate Creation: The Search for the Biological Origins of Sexual Orientation* (New York: Hyperion, 1996).

The best book on baking bread is Floss and Stan Dworkin, *Bake Your Own Bread and Be Healthier* (New York: Signet, 1972).

Prediction and Uncertainty

For the information on identical twins, including the example of one with and one without Turner's syndrome, see *Williams Obstetrics*, 16th ed., by Jack A. Pritchard and Paul C. MacDonald (New York: Appleton Century Crofts, 1980), p. 644.

For the example of type I diabetes, see Chandler Burr, *A Separate Creation*, p. 220.

For the comparison of biology with physics regarding uncertainty, see Robert Pollack, *Signs of Life: The Language and Meanings of DNA* (Boston: Houghton Mifflin, 1994). The quotes are from page 150.

On Authority

The "Madame Rosa" cartoon is by Nick Downes, printed in *Science* 238 (9 November 1987), p. 772, and thanks to Nelkin and Lindee for reprinting it in *The DNA Mystique* so I could find a reference for the copy yellowing on my wall.

Ruth Hubbard has repeatedly and forcefully made the point that prediction is not explanation, and knowledge does not inevitably yield cure. She uses the example of sickle-cell disease. For a fuller discussion, see Hubbard and Wald, *Exploding the Gene Myth*.

For a history of the ELSI project, see Robert Cook-Deegan, *The Gene Wars: Science, Politics and the Human Genome* (New York: Norton, 1994).

I want to thank John H. Evans for unpublished papers he was kind enough to share with me, and I look forward to the publication of his dissertation as a book. I drew here on a paper he presented to the Eastern Sociological Association meetings in 1997, "Playing God? Human Genetic

Engineering and the Transformation of Bioethical Debate, 1958—1995." The discussion of Esperanto is drawn from Jeffrey Stout, *Ethics after Babel: The Languages of Morals and Their Discontents* (Boston: Beacon, 1988). The report Evans analyzed was The President's Commission for the Study of Ethical Problems in Medicine and Biomedical and Behavioral Research, *Splicing Life: A Report on the Social and Ethical Issues of Genetic Engineering with Human Beings* (Washington, D.C.: U.S. Government Printing Office, 1982).

The report on Scottish focus groups on genetics is by Anne Kerr, Sarah Cunningham-Burley and Amanda Amos, "The New Genetics and Health: Exploring Lay Perceptions," a paper presented to the British Sociological Association Medical Sociology Conference, University of Edinborough, 1996.

The quote from Duster is in the preface of *Backdoor to Eugenics*, p. viii.

The "imagination" I have in mind is the one C. Wright Mills named in the title of his book, *The Sociological Imagination* (New York: Oxford University Press, 1959).

MAPPING THE PAST: THE MACROEUGENICS OF RACE

Inventing It
This section draws heavily on Jonathan Marks, *Human Biodiversity: Genes, Race and History* (New York: Aldine de Gruyter, 1995).

Seeing It
Lawrence A. Hirschfield, *Race in the Making: Cognition, Culture and the Child's Construction of Human Kinds* (Cambridge, Mass.: MIT Press, 1996).

George A. Theodorson and Achilles G. Theodorson, *A Modern Dictionary of Sociology* (New York: Barnes and Noble, 1969).

John Edgar Wideman, *Fatheralong: A Meditation on Fathers and Sons and Society* (New York: Pantheon, 1994). The quote is from the preface, pp. xii—xiii.

The Science and Politics of Race
Terman is quoted in Hubbard and Wald, *Exploding the Gene Myth*, p. 16.

There are many excellent histories of the eugenics movement; some that I drew upon are Daniel J. Kevles, *In the Name of Eugenics: Genetics and the*

Uses of Human Heredity (Berkeley, Ca.: University of California Press, 1986); Jonathan Marks's discussion throughout his book *Human Biodiversity;* a book that explores specifically the ways that American eugenics influenced the development of Nazi policy, Stefan Kuhl, *The Nazi Connection: Eugenics, American Racism, and German National Socialism* (New York: Oxford University Press, 1994)—the quotes comparing U.S. and German social policy are from pp. 19 and 36; and Robert Jay Lifton and Eric Markusen, *The Genocidal Mentality: Nazi Holocaust and Nuclear Threat* (New York: Basic Books, 1990)—the quote is from page 57.

For a history of infanticide and eugenics in the United States, see Martin S. Pernick, *The Black Stork: Eugenics and the Death of "Defective" Babies in American Medicine and Motion Pictures since 1915* (New York: Oxford University Press, 1996).

The Bell Curve: Intelligence and Class Structure in American Life, by Richard J. Herrnstein and Charles Murray (New York: Free Press, 1994), is discussed in greater detail later.

Through a Crystal, Darkly

The Maori woman quoted is Ella Henry, the executive director of Greenpeace New Zealand, from *Maori Sovereignty: The Maori Perspective,* ed. Hineani Melbourne (Auckland, New Zealand: Hodder Moa Beckett Publishers Ltd., 1995), p. 15. I thank Charmain for sharing this book with me.

On the presumed racelessness of whites, see Wideman, *Fatheralong,* pp. xviii—xix.

For the quote on the "assembly" of children, see Pollack, *Signs of Life,* p. 44.

(American) Racism

The film is *Bill Crosby on Prejudice* (Santa Monica, Ca.: Pyramid Film and Video, 1971).

For Whom the Bell Curves

I took this wonderfully evocative title from a phrase used by Michael Bérubé in *Life as We Know It: A Father, a Family and an Exceptional Child* (New York: Pantheon, 1996), p. 222.

The special issue of *Discover: The World of Science* was November 1994, Vol. 15, no. 11.

I used the 1996 edition of *The Bell Curve*, with a new afterword by Charles Murray. Richard J. Herrnstein and Charles Murray, *The Bell Curve: Intelligence and Class Structure in American Life* (New York: Free Press, 1996). Among the volumes of published critiques, I particularly recommend Claude Fischer et al., eds., *Inequality by Design: Cracking the Bell Curve Myth* (Princeton, N.J.: Princeton University Press, 1996); and Joel L. Kincheloe, Shirley R. Steinberg and Aaron D. Gresson, eds., *Measured Lies: The Bell Curve Examined* (New York: St. Martins, 1996).

Ellis Cose, *The Rage of the Privileged Class* (New York: Harper Collins, 1993) discusses the racism confronting middle-class African Americans.

William Ryan, *Blaming the Victim* (New York: Pantheon, 1971).

Heather Dalmage, unpublished doctoral dissertation, City University of New York, 1995.

Rates and Races

The quote from Jonathan Marks about blood group variations is on page 130 of his *Human Biodiversity;* the quote about sorting blocks is on page 159. A discussion of sorting by various characteristics giving us very different racial groupings can be found in the special issue of *Discover* magazine, in an article by Jared Diamond, "Race Without Color," pp. 82—92.

The Human Genome Diversity Project

The argument for the Human Genome Diversity Project is presented in L. Luca Cavalli-Sforza, Paolo Menozzi and Alberto Plazza, *The History and Geography of Human Genes*, abr. ed. (Princeton, N.J.: Princeton University Press, 1994).

The conference at Stanford was organized by Joan H. Fujimura, the Henry R. Luce Professor of Biotechnology and Science, and was called "Genetics and the Human Genome Project: Where Scientific and Public Cultures Meet." It was held November 3—4, 1995.

The quote from Donna J. Haraway is from *Modest Witness @ Second Millenium: Female Man Meets Oncomouse* (New York: Routledge, 1997), pp. 250—51.

A *New York Times* article on Aaron's Y chromosome appeared on January 7, 1997, and I thank the four people who mailed it to me.

The quote from Marks on the history of European explorations is from page 178 of his *Human Biodiversity*.

The discussion of gene flow between "isolated" tribes and the misperceptions of earlier anthropologists was made by John Moore at the Stanford conference and in his paper "Racism, Ethnogenesis and the Human Genome Diversity Project."

WRITING THE BODY: THE GENETICS OF ILLNESS

A Tale of Two Diseases

The idea of eugenics as coming in a back door is Troy Duster's, from his book *Backdoor to Eugenics*, an excellent source for a detailed, nuanced discussion of screening programs for Tay-Sachs disease and sickle-cell anemia, with particular emphasis on the latter. Hubbard and Wald also give some of this history in *Exploding the Gene Myth*. The quote from Hubbard is on pp. 64—65.

The book based on the CBS television series is *The Control of Life: The Twenty-First Century*, by Fred Warshofsky (New York: Viking, 1969); the quote is from p. 23.

In this chapter and then throughout the book, I refer to my previous research on prenatal testing in my book *The Tentative Pregnancy*, originally published by Viking in 1986, and most recently reissued by Norton in 1993.

The Quest

For histories of the Huntington's story, see Jeff Lyon and Peter Gorner, *Altered Fates: Gene Therapy and the Retooling of Human Life* (New York: Norton, 1995). The oft-quoted line about finding a killer in Red Lodge Montana is from Nancy Wexler, "Clairvoyance and Caution: Repercussions from the Human Genome Project," in *The Code of Codes: Scientific and Social Issues in the Human Genome Project*, ed. Daniel J. Kevles and LeRoy Hood (Cambridge, Ma.: Harvard University Press, 1992), pp. 211—43; quote p. 221.

A good example of a similar quest tale is Daniel A. Pollen, *Hannah's Heirs: The Quest for the Genetic Origins of Alzheimer's Disease* (New York: Oxford University Press, 1993).

The Wexler family quest is written by Alice Wexler, *Mapping Fate: A Memoir of Family, Risk and Genetic Research* (Berkeley, Ca.: University of California Press, 1995). The quote of Milton Wexler telling his daughter about Huntington's disease is from p. 43.

From the Breast
C. Wright Mills distinguished between "trouble" and "issue" in *The Sociological Imagination*. The article by David Plotkin is "Breast Cancer: The Good News and the Bad News," *Atlantic Monthly*, June 1996. The recent book on the breast is by Marilyn Yalom, *A History of the Breast* (New York: Knopf, 1997).

Cancer as (Not) a Genetic Disease
The book on immunity is by Emily Martin, *Flexible Bodies: The Role of Immunity in American Culture from the Days of Polio to the Age of AIDS* (Boston: Beacon, 1994). Any discussion of cancer and tuberculosis owes a debt to Susan Sontag, *Illness as Metaphor* (New York: Vintage, 1979). My understanding of TB historically and today also relies on the unpublished work of Rachel Grob.

For a good introduction to the genetics of cancer see Hubbard and Wald, *Exploding the Gene Myth*.

Peter Conrad has been working on the issue of how genetics is presented in the news for some years now; I draw upon papers he has presented over the years, in particular, "Public Eyes and Private Genes: Historical Frames, News Constructions and Social Problems," in *Social Problems* 44: 2 (May 1997), pp. 139—54.

Understanding Cancer
The discussion of cancer as metaphor comes of course from the work of Susan Sontag. My understanding of cancer draws upon early and unpublished work I did with Marcia Storch, and I am indebted to her for many helpful discussions.

Early Detection

The discussion of cervical cancer draws on a paper presented at the Second Scientific Symposium of Einseideln, Switzerland, October 5—8, 1995, "Screening for Cancer of the Cervix Uteri: A Questionable Practice," by James McCormick of the Department of Community Health at Trinity College of the University of Dublin. The discussion of breast cancer screening draws again on the article by David Plotkin in the *Atlantic Monthly.*

Breast Cancer

Nancy Press has done thoughtful work, from an anthropological perspective, on genetic testing and breast cancer, and I have been much influenced by her thinking. The work by Halsted was reprinted in *CA: A Cancer Journal for Clinicians,* March/April 1973. A good example of the early feminist approach to breast cancer can be found in Rose Kushner, *Breast Cancer: A Personal History and an Investigative Report* (New York: Harcourt Brace Jovanovich, 1975). For a good review of the work on prostate cancer and the selling of the PSA, see *Healthfacts,* edited by Maryann Napoli and published by the Center for Medical Consumers, 1997.

Contradictions

For more data on cancer risks and rates, including geographic patterns, see Hubbard and Wald, *Exploding the Gene Myth.*

IMAGINING THE FUTURE: THE MICROEUGENICS OF PROCREATION

The Gates of Life

A good summary of gene therapy for nonscientists is provided by "Making Gene Therapy Work," a special report of *Scientific American,* June 1997. Two book-length summaries of work in gene therapy, also written for lay people, are Larry Thompson, *Correcting the Code: Inventing the Genetic Cure for the Human Body* (New York: Simon and Schuster, 1994), and Lyon and Gorner, *Altered Fates.*

The disability activist quoted was on Dutch television in the spring of 1995, and I thank Annemiek Cuppen for translating for me. For a good introduction to a sociological perspective on disability, see Irving Kenneth Zola, *Missing Pieces: A Chronicle of Living with a Disability* (Philadelphia: Temple University Press, 1982).

Spoiling the Pregnancy

My research on Dutch midwives was conducted during my stay as a Fulbright scholar at the University of Groningen in the Netherlands, and would not have been possible without the help of Tjeerd Tymstra. Words cannot express how much I owe him, and his family, too. Focus groups were organized by two Dutch colleagues, Eva Roelofson and Inge Kamerbeek, and conducted by them along with members of my qualitative methods seminar. All of the quotes are from focus groups conducted during June and July of 1995.

Get a Map

The Dutch student is Ira Van Keulen: I owe her deeply for many productive conversations. The Clinton quote appeared in the *New York Times*, "In His Own Words," September 26, 1996.

The man quoted regarding the XYY fetus was quoted in *The Tentative Pregnancy*.

Beyond Prenatal Diagnosis

The quotes from women's experiences with prenatal diagnosis are drawn from my book *The Tentative Pregnancy*.

I've heard George Annas make this comment about "Cabbage Patch" babies at various conferences over the years.

Now You Can Choose!

I first heard the comparison of sex selection to selective abortion for disability from Deborah Kaplan at a panel on "Genetic Screening and Engineering" at the 14th National Conference on Women and the Law, Washington, D.C., April 10, 1983. The radio talk-show discussion of the television anchor is presented in a fine documentary on prenatal diagnosis, *The Burden of Knowledge: Moral Dilemmas in Prenatal Testing*, by Wendy Conquest, Bob Drake and Deni Elliott, Duma Productions, PO Box 272, Hanover NH 03755, 1994.

Determination

Evelyn Fox Keller is quoted from *Refiguring Life*.

Chandler Burr makes the comparison between male sexual orientation and handedness in *A Separate Creation*.

The midwife quoted was interviewed as part of my Fulbright-sponsored research discussed earlier.

On Order

Portions of this chapter—and small portions of several other chapters as well—appear in a chapter I wrote called "On Order," in *Clones and Clones: Facts and Fantasies about Human Cloning*, ed. Cass R. Sunstein and Martha Nussbaum (New York: Norton, 1998).

Many years ago my good friend and colleague Roslyn Weinman pointed out the move from "quantity control" to "quality control" in childbearing.

I thank Maren Lockwood Carden for showing me where the bread example was headed: Wonderbread.

CONCLUSION

In Search of Imagination

The quote from Schroedinger is on p. 67 of Fox Keller's *Refiguring Life*. The other quotes from Fox Keller are on gendering fertilization, p. 39; and the idea of 'before time zero,' on p. 33.

The quotes from Robert Pollack are on pp. 44 and 33 of *Signs of Life*.

There are many texts explaining symbolic interactionist theory. I wrote a chapter on it, "Symbolic Interactionism," in *The Renascence of Sociological Theory: Classical and Contemporary*, ed. Henry Etzkowitz and Ronald M. Glassman (Itasca, Ill.: F.E. Peacock, 1991), pp. 151—66. The original text by George Herbert Mead is *Mind, Self and Society* (Chicago: University of Chicago Press, 1934); and the classic summary statement of this theory is Herbert Blumer's *Symbolic Interactionism* (Englewood Cliffs, N.J.: Prentice Hall, 1969).

There are many discussions of the history of ensoulment as a concept, including Beverly Wildung Harrison, *Our Right to Choose: Toward a New Ethic of Abortion* (Boston: Beacon Press, 1983); see especially chapter 5, "The History of Christian Teaching on Abortion Reconceived," pp. 122—24 and 130—32. See also Paul Ramsey, "The Morality of Abortion," in *Life or Death: Ethics and Options*, ed. Daniel Laddy (Seattle: University of Washington Press, 1971). And there is an interesting discussion of the nuances of a debate of competing theories of preformation in the seventeenth and eighteenth centuries, between those who believed that the embryo unfolded from the sperm and those who believed it unfolded from the ovary. See Clara Pinto-Correla, *The Ovary of Eve: Egg and Sperm and Preformation*, (Chicago: University of Chicago Press, 1997).

Existential Orphanhood and Identity Politics: On Jews, Dwarfs and Gays

Jacki Ann Clipsham made her presentation at a conference on "Women and Genetics" organized by Becky Holmes in the fall of 1996 in Zanesville, Ohio. The quote from Leslie Fiedler on dwarfs is from *Freaks: Myths and Images of the Secret Self* (New York: Anchor Books, 1978), p. 90. Ruth Ricker points out that a modern version of community-building occurs in Little People of America. The novels that I mention are Ursula Hegi's *Stones from the River* (New York: Simon and Schuster, 1997) and John Irving's *A Son of the Circus* (New York: Random House, 1994). A different but very comparable set of issues comes up in the deaf community. This is explored nicely in Leah Hager Cohen, *Train Go Sorry: Inside a Deaf World* (New York: Vintage Books, 1995).

I heard Quentin Crisp say this about gays as orphans at a reading at Housing Works Bookshop in New York City on June 24, 1997.

I quote Heather Dalmage from her unpublished dissertation, City University of New York, 1996.

After I Finish

Faith Ringgold said this at a presentation at Brooklyn Friends' School's African American evening, February 28, 1997. Donna J. Haraway discusses OncoMouse in *Modest Witness @ Second Millennium*, p. 79.

It was Gena Corea's book *The Mother Machine: Reproductive Technologies from Artificial Insemination to Artificial Wombs* (New York: Harper and Row, 1985) that made me stop eating beef.

The quote from Michael Bérubé's book *Life as We Know It* is from p. 249.

I first heard that wonderful phrase about "calling forth" a child from Caroline Whitbeck.

Index